122765

13⁰⁰

1999
15

ALIENS

ALIENS

Can We Make Contact with
Extraterrestrial Intelligence?

Andrew J. H. Clark & David H. Clark

FROMM INTERNATIONAL
NEW YORK

First Fromm International Edition, 1999

Copyright © Andrew J. H. Clark and David H. Clark

All rights reserved under International and Pan-American Copyright Conventions.
Published in the United States by Fromm International Publishing Corporation,
New York.

Library of Congress Cataloging-in-Publication Data

Clark, Andrew J. H.
    Aliens : can we make contact with extraterrestrial intelligence? /
Andrew J.H. Clark & David H. Clark.
        p.   cm.
    Includes bibliographical references and index.
    ISBN 0-88064-233-5
    1. Life on other planets.   I. Clark, David H.   II. Title.
QB54.C564 1999   99-25649
576.8'39—dc21   CIP

10 9 8 7 6 5 4 3 2 1
Manufactured in the United States of America

Acknowledgments

We wish to thank our editor Fred Jordan, executive director of Fromm International, and our agent Al Zuckerman of Writers House for their excellent guidance and advice. A special mention needs to be made of Sir William McCrea, friend and mentor, who stimulated our interest in the fascinating subject of extraterrestrial intelligence. We wish to thank the many people, too numerous to name individually, with whom we have discussed our ideas. Most of all we wish to thank our family, for all their encouragement and loving support during the long hours of writing.

—Andrew and David Clark
November 1998

Contents

Prologue

"It's life ... but not as we know it."

Star Trek

C asual surfers of the Internet, with many hours to spare, can undertake a giddy ride through cyberspace merely by typing the single word "aliens" into their World Wide Web browser. Would they be any the wiser after such an experience? (Surfing the Internet has been likened to taking a drink from a fire hydrant; you can get very wet—but remain thirsty. Those thirsting after legitimate knowledge on aliens are at risk of being drowned in Internet trivia.) If, by sheer good fortune, the casual surfers had settled on the home page of the SETI Institute (http//:www.seti-inst.edu), or one of the other excellent scientific sites, they most certainly would have found it an enlightening and pleasurable experience, with the "search for extraterrestrial intelligence" (SETI, rhymes with "yeti") explained in a rational, entertaining, and scientific way. Unfortunately, a more likely outcome, especially if the surfers had been guided to the most frequently visited "aliens" Internet sites, is that they wasted many hours flitting between the rampant ravings of a miscellany of the chronically misinformed, the maliciously mischievous, and the congenitally insane. The subject of intelligent life elsewhere in the universe, although in reality a subject of legitimate scientific inquiry, has for many misinformed individuals taken on cult proportions largely unconstrained by the boundaries of conventional science. This book seeks to describe in a responsible manner the likelihood of detecting "*extra*terrestrial *i*ntelligence" (ETI, again rhyming with "yeti"). It presents perceived scientific wisdom to counter the myths that so often embrace consideration of the subject. It is not necessary to have a background in science to appreciate the arguments. The only science needed will be explained as we go along, and we hope that SETI will be a worthwhile introduction to legitimate science for those new to the subject.

While most thinking people feel sufficiently confident and well informed to discuss issues relating to such topics as the use of new technology, abuse of the environment, or climate change, when it comes to the issue of life elsewhere in the universe the realms of fiction and fantasy too often hold sway. The subject is so cloaked in ignorance that the popular fictional television series *The X-Files* is still believed by many viewers to be factual—and some 4 million citizens of the United States claim to have been abducted by aliens! (Sharp insurance companies have latched onto this claim, offering generous recompense for such an unnerving experience, so long as irrefutable evidence of the abduction can be produced. The morality of this profiteering from human gullibility seems to have escaped legal scrutiny.) The Web site recording the experiences of supposed "alien abductees" is depressingly inane; each confused fantasy no doubt represents a sad cry for attention.

We have sought to develop a rational approach to the subject of ETI and aliens (the word frequently used to describe a possible visitation to Earth by ETI). There is a delightful logic in choosing an acronym for intelligent life elsewhere in the cosmos, ETI, that rhymes with "yeti," the elusive creature of Himalayan legend.

Some previous respectable writing on the subject of SETI has focused on a single moment of contact with ETI and its aftermath. We have kept our attention on the present: What are the prospects for detecting ETI, and how do we go about the search? Science has much to tell us about ETI and how best to seek it out.

Setting the Scene

When the NASA Viking spacecraft landed on Mars in 1976, it

was equipped with instruments to look for the existence of simple life on the red planet. A century before, the astronomer Percival Lowell had argued that the canals he believed he saw on the surface of Mars indicated the existence of a vast civilization on the planet, irrigating its dusty soil by bringing water from the icy wastes of the planet's polar regions. The pioneer of radio communication, Guglielmo Marconi, believed he had detected radio signals from this Martian population. Such was the stature of these two men that their claims were taken extremely seriously. But the canals were eventually found to be mere tricks of light, and the radio signals were merely interference from the Sun. These Martian civilizations were no more grounded in truth than were the Martians of H. G. Wells' *War of the Worlds*.

A mere decade prior to the landing of Viking, before orbiting space probes gave us a close-up view of Mars, some scientists still believed that dark patches visible on the surface of Mars might be vegetation. One edition of *National Geographic* magazine even carried cover art displaying the imagined flora of Mars (but strictly no fauna). However, by the time Viking was designed, it was clear that Mars's surface was a barren rocky desert and that the only real hope of life was in the form of simple microorganisms. The Viking lander carried three experiments designed to search for evidence of such simple life. A robot arm grabbed samples of soil, which were analyzed in various ways by onboard instruments. One of the three experiments, whose design was borrowed from a system developed to test the sanitation of bathrooms, provided a positive detection for the presence of microbes, but the other two (supposedly more sensitive) experiments failed to detect the presence of any organic molecules. The single positive result was discounted as spurious by the mission team; they chose to go with the two negative results. Some scientists, particularly those involved in the overruled experiment, felt that NASA's subsequent categorical pronouncement that there was no firm evidence of even simple life on Mars overstated the conclusions of the mission. They might just have been right.

In 1984 a potato-sized meteorite, later labeled *ALH 84001*, was discovered in Antarctica. (*ALH* identifies the location of the find; *84,* the year; and *001* indicates it was the first meteorite discovered that year.) Analysis of its chemical composition showed element abundances akin to those measured on Mars by the Viking lander mission. But how could this piece of Martian rock, estimated to be the order of 3.6 billion years old, have reached the Earth? A plausible scenario runs as follows: About 16 million years ago a huge asteroid or comet collided with Mars. Some debris from the impact was blasted into interplanetary space, drifting through the Solar System until the "potato" fragment crashed into Antarctica an estimated 13,000 years ago.

The interest in this particular meteorite was heightened in 1996 when electron microscope images showed what looked to be tiny bacteria-like structures deep inside it. The appearance of the structures is interesting enough; the fact that they are associated with complex organic molecules found on Earth when microorganisms die and decompose adds to the intrigue. Some believe that the Antarctic meteorite provides evidence of simple microbes on Mars some 3.6 billion years ago, when much of Mars was believed to have been covered with water and the planet was considerably warmer than now. It is thought that the conditions on primordial Mars would have been fairly similar to the conditions on primordial Earth when life originated in the oceans. It has even been suggested that conditions for the origin of life might have been marginally better on Mars than on Earth, and that the first primitive life-forms reached Earth on spallated rocks, like 84001, from Mars. Subsequently, it is surmised, Earth proved to be a fairer environment for evolution than Mars. Changes to the Martian environment could not sustain the advancement of life. On Earth they could. This hypothesis, that life on Earth originally came from Mars, while intriguing, remains somewhat controversial and unproved.

Not surprisingly, NASA sought maximum publicity for the discovery of the "Mars bugs." President Clinton announced, "Today Rock 84001 speaks to us across those billions of years and millions of miles. It speaks of the possibility of life. If this discovery is confirmed it will surely be one of the most stunning insights into our universe that science has ever uncovered." The late Carl Sagan, noted planetary scientist and author, was even more poetic:

"If the results are verified it is a turning point in human history, suggesting that life exists not just on two planets in one paltry solar system but throughout this magnificent universe."

Good news for those who believe that intelligent life will eventually be found elsewhere in the universe? Possibly. But the reality of Rock 84001 microbes remains uncertain, many scientists believing that the fossil microbe-shaped features are merely mineral crystals, and others raising the prospect of contamination with organic molecules of earthly origin. The issue is likely to remain open until planned future missions to Mars perform studies using instruments more sensitive than the Viking ones were.

The possibility that life might arise spontaneously through commonplace chemical reactions (in the presence of water and radiation from a Sun-like star) is now accepted by most scientists. From this perspective, conclusive evidence that primitive life *never* formed on Mars could be more surprising than the present tantalizing positive evidence. But what additional factors are needed to allow simple microbes, which present scientific understanding suggests could be infinitely replicated across the universe, to evolve to intelligent life-forms with consciousness, able to contemplate their origins and to seek out other intelligent life-forms elsewhere in the cosmos? There might well be plenty of "slime" out there (simple *life* in a *microbial* ecosystem); but what special factors allow that "slime" to evolve to ETI, and does ETI wish to communicate its existence to other intelligent beings (including us)?

The "Mars bugs" saga brought renewed popular interest in the possibility of extraterrestrial life. The media worked itself into a frenzy. All this was great publicity for NASA, chasing congressional support for a new series of planetary missions. Our own interest in life elsewhere in the cosmos had been stimulated somewhat earlier than the Mars bugs excitement.

A Letter of Particular Interest

In July 1977, one of us (D. H. C.) received a letter from a much-valued professional colleague, Sir William McCrea.

McCrea is one of the great astronomers of the twentieth century. His letter started as follows:

> Dear David,
>
> Do you ever think about LIFE? Lots of people do, but I think more and more that a particular astronomical aspect could be isolated and discussed usefully now, and followed up for a goodish time to come.
>
> Briefly the problem is, supposing that life in some elementary form that we know about is available on every planet in the Galaxy, what is the chance that creatures like humans will evolve...?
>
> In a way it is an obvious problem. But the constraints are important. It does not ask about the origin of life. It does not ask if life has to originate on a planet. It does not ask about whether very different forms of "life" are possible. As stated the problem is about life like ours on a planet.

McCrea then went on to argue in his intriguing five-page letter that certain ranges of various parameters, such as gravity, the constituency of an atmosphere, incident radiation, surface temperature, the existence of water, and so forth, would have to be considered for any planet sustaining human-like life. McCrea had approached the understanding of a major mystery in a clear, logical manner. It needs to be stressed that McCrea was not talking about "creatures like humans" in the anatomical sense, but in the sense of intelligent life-forms that could develop an advanced technology which would enable them to communicate across the vast expanse of the cosmos. He was talking about ETI, *not* "humanoids." He proposed a new research effort, to establish the likely emergence of "creatures like humans" working from fundamental scientific principles. With a characteristic courtesy, he ended his letter, "Don't let this worry you and please say if it all seems not worthwhile. With my best wishes, Bill."

This letter from McCrea has acted as a stimulus to our own interest in the origin of intelligent life, and the possibility of detecting ETI. It is a field of research that has intrigued scholars over the centuries, and now engages some of the top researchers of the modern era.

Despite the hesitation McCrea displayed at the end of his letter of over 20 years ago, the studies he suggested then remain eminently worthwhile today. Evidence has been accumulating over recent years that, as long supposed, planetary systems are a frequent occurrence for stars—indeed perhaps the majority of stars have planets. (McCrea was a pioneer of theories of planetary formation, and always argued that planets were a natural consequence of the formation of most stars.) The fundamental molecular building blocks for life systems are detected in astronomical observations, and it is now suspected that these could be linked with relative ease into self-replicating forms on any hot young planet. It is accepted that some simple life-form is bound to evolve wherever the conditions are similar to those that existed on planet Earth 4 billion years ago when single-cell creatures first appeared.

If simple single-cell systems are commonplace throughout the cosmos, a problem is to establish how often these could evolve to intelligent life-forms. Increasingly scientists ask, might intelligent life-forms on other planetary systems be attempting to signal their existence? The full power of technology has been applied to the challenge in recent decades. In 1960 U.S. astronomer Frank Drake initiated the first SETI program. The Soviets also started a program to search for radio transmissions from highly advanced ETI. Although SETI funding from U.S. federal sources has now been discontinued by the Congress, the U.S. program still attracts private funding. The heavens are scanned by radio telescopes, listening for evidence of artificial signals from extraterrestrial civilizations that have achieved a technological capability equal to or exceeding our own. Nothing has been detected so far, but SETI surveys continue apace.

There are skeptics who would argue that since almost four decades of diligent search have failed to make a single certain detection, then ETI cannot possibly exist. Not so. What the failure to make a detection does emphasize is that the search for ETI will be difficult and long; it is one of the most important challenges facing contemporary science. SETI was never going to be easy.

It needs to be appreciated that SETI involves listening for signals from other intelligent beings. It is not about transmitting signals revealing our existence (although some such transmissions have been made). It is not about engaging in conversation with ETI (a near-impossible task over the vast distances of the cosmos). We have to imagine the interests and motivations of civilizations that are likely to be spectacularly more advanced than our own, and assume that they want to advertise their existence and communicate their thoughts to us. It is important that we remain alert and listen.

There are some pretty big assumptions involved in all of this: plenty of "ifs and buts" and a wealth of "maybes." However with a few initial leaps in imagination, it is possible to establish a very sound basis for a search for ETI, as we hope to demonstrate in this book.

If highly advanced civilizations exist elsewhere in the cosmos, they might have developed a space travel capability spectacularly more sophisticated than our own. The great Italian scientist Enrico Fermi pondered the implications of this while engaged in the Manhattan Project to build the first atom bomb during World War II. Fermi was convinced that there must be multitudes of advanced civilizations in the cosmos, many with the technology for space travel. He then asked colleagues, "Where are they?" The Fermi question can be asked in a slightly different way: If they are there, why aren't they here? (Fermi's Hungarian-born colleague is reported to have said, "They are among us, but they call themselves Hungarians.")

In early 1997, commenting on the reported discovery of the microbes in the meteorite sample from Mars, astrophysicist Stephen Hawking caught the spirit of the Fermi question when he noted:

> I see no reason to believe life cannot develop elsewhere, but the fact that we do not seem to have been visited by aliens needs an explanation.
>
> Meeting a more advanced civilization might be a bit like the original inhabitants of America meeting Columbus, I don't think they were better off for it.

One can derive from the same widely held scientific beliefs that form the basis for SETI experiments the surprising consequence that the Earth should perhaps have already been visited by alien extraterrestrial life-forms. Despite the oft-quoted sightings of "flying saucers," there is no sound scientific evidence of an alien visitation. Why is it that this does not seem to have happened, especially since the case for ETI proffered by the SETI community appears to be so soundly based? And should we fear that a visit might well be overdue?

The Three Big Questions

We have found it convenient to structure this book around three big questions that we claim need to be addressed when looking at the topic of ETI. The answers to the three questions, it turns out, will lead us to very similar conclusions about the occurrence and nature of ETI. The three questions are as follows:

1. The SETI question: If intelligent life-forms are in fact commonplace in the Galaxy, as many eminent researchers now claim, why has the power of modern technology and years of dedicated research failed to detect signals from them?

   The nondetection of SETI signals can now place limits on the occurrence of intelligent extraterrestrial life. However, absence of evidence should not be interpreted as evidence of absence. All that the SETI results show us is that the number of advanced civilizations may be significantly lower than the more optimistic estimates made in the past, or that the intelligent civilizations may be reluctant or unable to make contact, or that the techniques used so far have not matched the preferred method of communication of extraterrestrials, or that a myriad of other possible reasons for nondetection may apply.

   The SETI pioneers are well aware that they are seeking out a very tiny needle in a very large cosmic haystack!

2. The McCrea question: If life in some elementary form that we know about is available on every planet in the cosmos, what

is the chance that creatures like humans will evolve else-where?

Astronomical observations from the past decade enable informed estimates to be made of the number of planetary systems in the cosmos with conditions that could have resem-bled those on the nascent planet Earth and that could subsequently sustain the evolution of life. Simple life may be simple to form. But the path to an intelligent civilization with an advanced technology allowing communication by radio or some other means is exceedingly complex, if the evolution of humans is taken as a guide.

3. The Fermi question: If they are there, why aren't they here?

The likely incidence of ETI that is implied by the SETI question, and the tight limits that are required by the McCrea question, as we will discover later, do not preclude possible visits from aliens. There are major uncertainties surrounding such matters as energy requirements, the economics of inter-stellar travel, and the possible motivation of ETI to strike out for other star systems. But one certainly cannot say, on scien-tific grounds, that visits from extraterrestrials spectacularly more technically advanced than humans are not possible. However, the distances involved in interstellar travel mean that even if alien spacecraft were setting out toward planet Earth with reasonable frequency, many thousands of years might often pass between visits. The frequent claims of encounters with aliens (popularized in the tabloid newspa-pers) can largely be dismissed as fantasy and mischief, although hidden among all the spurious reports, there are cer-tain to be some sightings of unusual phenomena worthy of serious scientific consideration. Aliens they might not be; gen-uine phenomena deserving scientific explanations they could well be.

The fact that there is no conclusive evidence of alien arrival during the period of recorded history should not be construed as meaning that ETI does not exist, or does not have the technology for interstellar travel; it may merely mean that we are now overdue for a visitation. It is a great pity that serious scientific consideration of this possibility has not been undertaken—in large measure because the topic of alien visitation has been hijacked by the pseudoscience fringe.

Genesis

One of the monumental achievements of twentieth-century science has been to gain an excellent understanding of how the universe was formed in a "big bang" some 15 billion years ago—and what its ultimate fate might be. The big bang is envisaged to be the epoch of creation. The early universe was composed of 75 percent hydrogen and 25 percent helium, plus radiation. As the nascent universe expanded, galaxies were formed, and within the galaxies the first generations of stars and planets were born. Although the current theories for the formation of galaxies leave some unanswered questions, the general picture is understood. Scientists can now describe the birth, life, and death of stars and planetary systems with some confidence. It is in the stars (giant globes of gas at extreme temperatures) where the elements that make up everything of familiar experience (including the carbon, nitrogen, and oxygen that combine in the building blocks of life-forms) are forged. Hence, ancient mythologies that described humans as being "the children of the stars" contained a semblance of truth! It is a sobering thought that all the elements in our bodies, other than hydrogen, have been processed through the stars.

Humans through the millennia have viewed the heavens with wonder and with awe, sensing the vastness of space, the power of the creation, and perhaps even something of their own origins as they looked out into a clear night sky. Until the past half-century, however, they could have had no real appreciation of the true enormity of the cosmos, its cataclysmic origin, or their own close relationship to the stars. Our ancestors even wondered whether they might not be alone in the universe. Metrodorus, a fourth-century-B.C. philosopher, had no doubts: "To consider the Earth the only populated world in infinite space is as absurd as to assert that in an entire field sown with millet only one grain will grow." However, opinion on the possibility of other populated worlds was divided, and in recent centuries the prospect caused anguish within evolving religious doctrine. Giordano Bruno was burned alive at the stake in 1600 by the Roman Inquisition for his cosmological views, including his belief that many populated worlds existed. One of the more powerfully expressed opinions was that of Thomas Paine in his *Age of Reason* (1793), where he stated:

> To believe that God created a plurality of worlds at least as numerous as what we call stars, renders the Christian system of faith at once little and ridiculous and scatters it in the mind like feathers in the air. The two beliefs cannot be held together in the same mind; and he who thinks that he believes in both has thought but little of either.

Modern scientific understanding, and the liberalization of traditional religious dogma, now allow the two beliefs to be reconciled.

The scope of present-day astronomical research extends from the origin of the universe, beyond its present turbulent state, to speculation about its ultimate fate. It extends from the Earth's nearest, and comparatively well-understood planetary and stellar neighbors, to bizarre and enigmatic objects at the extremities of the observable universe. It now actively embraces the prospect of ETI, and diligently seeks it out. SETI has secured a legitimate place in modern astronomical research, and uses major research facilities to look for signals from extraterrestrials. Only a few of the world's largest radio telescopes have not searched for ETI at some time.

Because of the vast distances involved in the cosmos, it is no longer appropriate to use familiar units of distance such as the mile or kilometer. Instead astronomers use the *light-year*, the distance a pulse of light travels in 1 year. Since the speed of light is a staggering 300,000 kilometers each second, a light-year is a considerable distance—some 10 trillion kilometers. (*Billion* is the term used for 1,000 million, and *trillion* is the term used for a million million.) At 300,000 kilometers per second, the light from the Sun takes about 8 minutes to reach the Earth. The Solar System is about 12 *light-hours* in diameter. The nearest star to the Sun is almost 5 light-years away. The extreme distances to the stars present particular difficulties for SETI, as we will discover.

To describe the vast distances in the cosmos and the long time intervals involved in the evolution of the universe, scientists find it convenient to use a system of "powers of ten." Remember how in the story *Alice's Adventures in Wonderland,* Alice grew larger when she ate the cake? Let us imagine that one bite of cake meant increasing in size ten times—and each successive bite would mean a further tenfold increase in size. After the first bite, a 1-meter-high Alice would be 10 meters tall; after two bites, she would be 100 meters tall; and after three bites, she would be 1,000 meters high. By the fourth bite her head would be at the cruising altitude of a jumbo jet, and by the sixth bite she would be well on the way to the Moon. This process demonstrates powers of ten; the second imaginary bite is ten to the power of two (100 meters), written as 10^2 meters, and the fourth imaginary bite is ten to the power of four (10,000 meters), written 10^4 meters, and so forth. To get to the distances of the stars requires sixteen imaginary bites of cake, and to get to the extremities of the cosmos would require twenty-five imaginary bites of cake (that is, we are at distances of 10^{25} meters). The number 10^{25} is very large—it is the number one followed by twenty-five zeros (10,000,000,000,000,000,000,000,000 meters). A light-year is equivalent to 10^{16} meters. The value 10^3 is often summarized by the prefix *kilo* (thus, a kilometer is 10^3 meters); 10^6, by the prefix *mega*; 10^9, by *giga*; and 10^{12}, by *tera*. A million is 10^6; a billion is 10^9; and a trillion is 10^{12}. Astronomy has to deal with lots of very large numbers, and the use of powers of ten certainly saves on ink and paper in writing out endless strings of zeros. But despite its convenience, powers of ten can disguise the enormous distances and time spans we need to deal with in discussing the search for ETI. Think big—and if it helps to think big, try to imagine the long string of zeros.

Stars are not uniformly scattered throughout the cosmos, but accumulate in vast conglomerates called *galaxies*, containing many billions of stars. Galaxies themselves tend to accumulate in clusters. Our Sun is just one of an estimated 400 billion (4×10^{11}) stars within our Galaxy (the capital letter signifying by convention the galaxy the Solar System lies in, rather than any old galaxy). Our Galaxy is called the *Milky Way* (from the appearance of the nebulous band of neighboring stars stretching across the night sky). The Milky Way Galaxy is discus shaped, a full 100,000 (10^5) light-years across at its widest, with our Sun occupying a rather insignificant location closer to its periphery than its heart. The bright stars of the Milky Way lie in intertwined spiral arms. Such spiral formations are a common species of galaxy, although many galaxies adopt a more amorphous elliptical shape or an irregular shape. A tenuous interstellar medium lies between the stars; space is not quite a perfect vacuum, although it comes close to it (three grains of sand in Yankee Stadium are more closely packed than atoms in the interstellar medium).

If our place within the Milky Way Galaxy seems insignificant, then the place of the Milky Way within the universe seems even more so. The observable universe is believed to contain at least 10 billion (10^{10}) galaxies, clustering together in the thousands but still spaced from one another by millions of light-years. The Milky Way lies within a cluster of galaxies called the *Local Group*.

If the vast distances in the cosmos are difficult to assimilate, further adjustments to our terrestrial scale of thinking are required if we are to appreciate the mass scales involved in the universe. Scientists choose to measure the mass of objects of common experience in terms of a convenient standard, the kilogram. (There are about 2.2 pounds in a kilogram.) Thus, for example, an adult male may have a mass of about 80 kilograms. The mass of planet Earth is 6 trillion, trillion (10^{24}) kilograms! The Sun is some 300,000 times more massive than the Earth. The mass of the Milky Way is probably at least 500 billion (5×10^{11}) times that of the Sun! And the mass of the universe? Well, it is certainly greater (perhaps very much greater) than a billion, trillion (10^{21}) solar masses! Just how much matter there is in the universe remains somewhat uncertain, and will determine its ultimate destiny.

No less of a challenge to the human imagination are the time scales involved in describing astronomical phenomena. Earthbound events are conveniently measured in terms of the sidereal year, the time it takes the Earth to complete one orbit about the Sun, measured relative to the fixed stars. The Sun and other stars orbit around the center of the Milky Way, like a gigantic Catherine wheel. At the Sun's distance from the center, the stars take over 200 million years to complete one revolution. The Sun is believed to be some 5 billion years old, and will survive for a similar period. The universe itself is thought to be some 15 billion years old. Time is not a problem when it comes to thinking about the eons needed for the processing of elements in the stars, to enable the eventual formation of planetary systems, and then for life to develop. Time is something the universe has had plenty of, and has plenty more to come. There has certainly been no shortage of time for any ETI to evolve. After all, the path of evolution on Earth led from primitive sea creatures to humans in approximately 500 million years. Just try to imagine what several billion years of evolution to ETI might produce! When it comes to the question of whether extraterrestrial life-forms could evolve to levels of intelligence and gain technological capabilities far beyond our own, time is not an issue. The universe has provided time aplenty.

In thinking about the cosmos, one needs to get used to *very* big numbers for times, sizes, and distances!

When looking out into the cosmos through a telescope, because of the finite speed of light, we are seeing not only deep into space, but also back in time. Thus, the nearby stars are viewed as they were several years ago. More distant stars within the Milky Way are seen as they were thousands of years ago when the light now reaching the Earth commenced its cosmic journey. The nearby galaxies appear as they were millions of years ago, and the more distant galaxies as they were hundreds or even thousands of millions of years ago. Many of the objects we observe do not still exist at this instant, at least in the form we presently see them. Thus, the history of the universe is laid out for Earth-bound heaven gazers to contemplate. The telescope represents a form of time machine in which we can study stars and galaxies at various stages of their evolution: nascent stars procreated from giant clouds of interstellar gas and dust, young stars, old stars, dying stars, and dead stars—young galaxies, interacting galaxies, and galaxies being torn apart. The universe reveals itself as a spectacle of unfolding drama, as stars and star systems are born and die, often violently.

On the universal scale, planet Earth must be considered to be no more than a mere speck of cosmic sand. Although pre-Renaissance theology placed the Earth and its peoples at the center of God's creation, we must now accept a more humble place in the grand scheme of things. The Milky Way Galaxy is not special; the Sun is not special; the Earth is not special; humans are not a special life-form.

Is it really conceivable that a cosmos of such vastness could have produced its single intelligent life-form on a planetary system of little consequence, circling a star of common type, on the outskirts of a very ordinary galaxy, within a cluster of galaxies of no special character? The sheer ordinariness of planet Earth challenges any assumption that humans could possibly be unique as an intelligent life-form.

Don't Be Fooled by Large Numbers

The problem with the statement in the previous paragraph is that it assumes that if there are enough galaxies, with enough stars with planets, and loads of time, then intelligent life-forms are sure to show up. It is the same "logic" used by the compulsive gambler who keeps buying lottery tickets convinced that an eventual win is certain. But we cannot be certain how likely the emergence of complex life is. We do know that matter appears to have a tendency to form more and more complex structures. However, we will show later that the evolutionary path leading to humans has been the fortuitous outcome of thousands of random events so unlikely that even in a vast cosmos the evolution of creatures with the capabilities of humans to develop advanced technology may be very rare indeed. Humans only just made it; any ETI is likely to face similar evolutionary uncertainties.

We cannot hope to find the answers to the SETI, McCrea, and Fermi questions unless we look for evidence of simple life-forms on other planets in our Solar System, search for life-bearing planets around distant stars, and listen for radio signals from ETI. If we do not search, we will not find. The search might be long and complex, and the outcome unknown, but the search represents one of the great challenges for humankind. Can there be any more important question to help us to appreciate the true place of humans in the cosmos than, Where is ETI?

PART A

The SETI Question

•

If intelligent life-forms are in fact commonplace in the Galaxy, as many eminent researchers now claim, why has the power of modern technology and years of dedicated research failed to detect signals from them?

1. The Drake Equation

"It is a capital mistake to theorise before one has data."

Conan Doyle

It is widely believed in the world of publishing (although hopefully mistakenly so) that the inclusion of a single mathematical equation in a book for a general readership will slash the potential sales. Therefore, the inclusion in this book of a whole chapter on an equation might appear to be the ultimate literary folly. However, the so-called Drake equation is at the heart of the modern search for extraterrestrial intelligence (SETI), and hence an understanding of what the equation is about will prove extremely beneficial in addressing the issues surrounding the possible emergence of intelligent life in the cosmos. One really cannot understand SETI without trying to understand what the Drake equation is telling us. It is no good hiding it away in an appendix, or pretending it does not exist. If you want to appreciate the challenges of SETI, you have to understand what the Drake equation is all about. It is as simple as that. Hence, we have decided to be brave, and to put the equation right up front. No ambiguity—only clear explanations. No fudging the issues—just straight talking. No lame excuses—just the hope that the conventional publishing wisdom about equations is wrong. Please stay with us; the effort will be worth it, we promise. Here goes.

The Drake equation has become one of the icons of modern science, almost on a par with Einstein's $E = mc^2$. (You might even see it displayed on a T-shirt!) Don't be put off by the mathematics in what follows—the Drake equation is really just about common sense. It is not necessary to try to understand the mathematics fully. If you find equations confusing, read over them and concentrate on the words of explanation. We promise that the effort will have its rewards in understanding the challenges of SETI, and appreciating the likelihood that SETI will ultimately be successful.

An Imaginary Drake Equation

To understand the purpose of the Drake equation, let us imagine a situation somewhat more familiar to everyday experience than the search for ETI, namely, buying a pair of new shoes. We could have chosen just about anything that people frequently have to make choices about as a suitable illustration, but shoes seem to be suitably gender and age neutral. Suppose you saw a notice outside a shoe store grandly announcing, "Over 12,000 pairs of shoes to choose from!" If you were in need of a new pair of shoes, a claim of such a large stock would undoubtedly impress you. You might imagine that you would have ample choice from such an impressive number of pairs of shoes, on the assumption that the owner had established the stock evenly across all possible sizes and styles. Think again! The choice may not be as great as you at first suppose.

Of the 12,000 shoes, let us assume that half are for men and half are for women. Your choice is immediately reduced to 6,000. Let us then assume that shoes come in ten sizes. Now your choice is down to 600. Let us assume that there are three standard widths for each size. Now your choice is down to 200. We should then concentrate on the four basic shoe types available: leather upper and sole, lace-up; leather upper and sole, slip-on; leather upper and composite sole, lace-up; leather upper and composite sole, slip-on. By opting for one of these types, your choice of shoes is now down to 50. Then suppose there are five standard colors: black, dark brown, tan, gray, and white. Now the choice is down to 10. If you went into the shoe store expecting a very wide choice, you would soon realize otherwise:

After you state your foot size, width, sole, and color preference, the store assistant may produce only ten pairs that you can then consider for style, comfort, and favorite brand name. This is not such a great selection after all, should you have been attracted initially by the sign signaling a massive stock. The shoe store owner had sought to attract your attention by appealing to large numbers—12,000 shoes in stock. Essentially the unstated claim was "with so many shoes in stock, is it really conceivable that we won't have a pair suited to your needs?" Remember the assertion of Metrodorus: "To consider the Earth the only populated world in infinite space is as absurd as to assert that in an entire field sown with millet only one grain will grow." The shoe store owner is saying, in effect, "to consider that we will not have your choice of shoes in such a large stock is as absurd as to assert that in an entire field sown with millet only one grain will grow."

Like the shoe store owner, some of the proponents of SETI love appealing to the large numbers readily available in astronomy to attract your attention: Is it really conceivable that in a universe containing billions of galaxies—with each galaxy containing billions of stars—humans are the only intelligent life-form to have evolved? The answer may be no, but do not be fooled by the astronomically large numbers into expecting a selection of intelligent beings any greater than your disappointingly small choice of shoes from a massive stock.

To establish an equation to represent our choice of shoes, we need to introduce a form of mathematical shorthand. For the number of shoes we are likely to have available for a final choice, let us assign a symbol N. (Using symbols in mathematics just saves writing a lot of words: for "the number of shoes available for a final choice," merely think "N.") Suppose the total number of shoes in the store is N^*. For the fraction of shoes available for either gender, we will use a symbol f_g; the fraction of a particular size will be given the symbol f_s; the fraction of a particular width will be f_w; the fraction of a particular sole-upper combination will be f_u; and the fraction of a particular color will be f_c. We can then construct a very simple equation, where each symbol following the equal sign (=) is multiplied by the one that follows. The equation is as follows:

$$N = N^{*}f_{g}f_{s}f_{w}f_{u}f_{c}$$

Each of the f symbols acts as a "filter" on our choice; we filter out the sizes we do not want, we filter out the widths we do not want, and so forth. For the values assumed in the above description, we would have the following:

$$N = 12{,}000 \times 0.5 \times 0.1 \times 0.33 \times 0.25 \times 0.2 = 10$$

(The identification of gender means that the total number is multiplied by 0.5; the selection from ten sizes means we multiply again by 0.1; the selection from three widths means we multiply again by 0.33; the selection from four sole types means we multiply again by 0.25; and the selection from five colors means we multiply finally by 0.2, giving the final number of possibly suitable shoes as just 10 out of the stock of 12,000.)

Just for fun, we will call our equation "the Drake equation for shoes." Using a form of mathematical shorthand, we have started with the total number of shoes in the shop, and estimated how many might be suitable for purchase; not many, it turns out, from such an impressively large stock. Do not worry too much about the use of symbols in the "Drake equation for shoes"; just try to appreciate the method of estimation being used.

Here we are merely using the tools of mathematics to present a point of view in a logical and consistent way. But it is not the tools of mathematics that are important for our purpose here; it is the "common sense" that counts. And common sense (as well as everyday experience) soon leads to the conclusion that it is often difficult to find the shoes of your choice when you go shopping for them, regardless of how large the initial stock in any shop might be. Our imagined shoe store owner's appeal to large numbers has actually disguised a somewhat modest choice.

Our "Drake equation for shoes" is the product of "probabilities"; that is, what is the probability that a pair of shoes will be the correct size, what is the probability that they will be the correct width, what is the probability that they will be the right sole type, and what is the probability that they will be the desired color? In the imaginary situation we considered, the product of probabilities gives an overall probability that from a stock of 12,000 shoes, only about 10 would be of interest to any of us.

The overall probability of a SETI detection also depends on multiplying several individual (but interrelated) probabilities.

The True Drake Equation

The true Drake equation used for SETI has a similar form to our imaginary "Drake equation for shoes." But instead of being impressed by the sign outside the shoe store announcing "Over 12,000 pairs of shoes to choose from!" we are looking for life in a fabulously stocked Milky Way Galaxy where we have "Over 400 billion stars to choose from!" Again, as with the shoes, the choice looks enormously impressive. But again, as with the shoes, once we start applying filters to get rid of the stars and planetary systems that do not fit our requirements, we are left with a somewhat more modest range of planetary systems on which intelligent life might have evolved. We must resist the temptation to be overly impressed by any appeal to large numbers!

The true Drake equation comes in a variety of forms, a convenient one for comparison with our shoe equation being as follows:

$$N = N^* f_p n_e f_l f_i f_c f_L$$

Here N is the number of presently existing advanced civilizations in the Milky Way Galaxy able to communicate by radio transmissions, the number we would like to estimate (albeit with recognized uncertainty) to establish a basis for SETI. N^* is the number of stars in the Milky Way Galaxy of the type that might survive long enough for life to evolve on any planetary system. The symbol f_p is the fraction of those stars with planets. The symbol n_e is the average number of planets per solar system whose environments are suitable for life. The symbol f_l is the fraction of habitable planets on which life occurs. The symbol f_i is the fraction of life-bearing planets on which intelligence evolves. The symbol f_c is the fraction of planets with intelligence where radio communication potential develops. The symbol f_L is the fraction of advanced civilizations existing at this time.

N^*, f_p, and n_e are *astronomical factors*—it is the astronomers who can provide reasonable estimates of what these might be. By contrast, the terms f_l and f_i are the *biological factors*, where we require guidance from evolutionary biologists. And finally, the *sociological factors* f_c and f_L require some imaginative thinking about the social evolution of intelligent civilizations. The three astronomical factors lend themselves to observational investigation (and are likely to cause the least dispute). The two biological factors can reflect evolutionary theory (although here there is some dispute about this approach). However, for the two sociological factors we are into the realms of unbridled speculation, which does leave a feeling of unease.

It is fair to observe that, complicated as the Drake equation might already appear, many additional (albeit possibly less important) astronomical, biological, and sociological terms could be added to make it appear more robust to counterargument. For example, the impact of various terrestrial (for example, volcanoes) and extraterrestrial (for example, cometary impact) influences on the evolution of life on Earth could be considered. Later, we will see just how important such influences can be. A more rigorous Drake equation could accommodate many such additional factors, although this might divert attention from its primary purpose of focusing consideration on the principal factors that could determine the conditions for ETI. In Chapter 3 we will revisit this matter, and look at the effect of including plausible additional factors. We will ignore this potential complication until then.

The main appeal of the form of the Drake equation used here is clearly the way in which each successive term acts as a filter through which only those star systems we might potentially be interested in pass. We are then left with the residue of advanced civilizations, which in the case of contemporary SETI is usually understood to mean civilizations capable of interstellar communication.

The Start of It All

The SETI bandwagon was set rolling, at least in a public way, in a famous paper written by Giuseppe Cocconi and Philip Morrison, and published in the scientific journal *Nature* in 1959. In this classic work they argued that radio signals from ETI might be detectable with the radio technology then available. (Morrison, a gifted physicist and a great popularizer of science, was one of the Manhattan Project pioneers. His book reviews in *Scientific American* achieved cult status in recent decades. Cocconi was a student of the great Enrico Fermi.) In a rousing final paragraph to their fine paper "Searching for Interstellar Communications," the two Cornell University scientists wrote:

> The reader may seek to consign these speculations wholly to the domain of science fiction. We submit, rather, that the foregoing line of argument demonstrates that the presence of interstellar signals is entirely consistent with all we now know, and that if signals are present the means of detecting them is now at hand. Few will deny the profound importance, practical and philosophical, which the detection of interstellar communications would have. We therefore feel that a discriminating search for signals deserves a considerable effort. The probability of success is difficult to estimate; but if we never search, the chance of success is zero.

Radio astronomy is at the heart of SETI. Radio waves were discovered by Heinrich Hertz in 1887. The best-known early pioneer of radio communication was Guglielmo Marconi. He predicted that radio would eventually be used for communicating with intelligence elsewhere in the cosmos. Radio astronomy had its beginnings in the experiments of a Bell Telephone Laboratories engineer, Karl Jansky, during the 1930s. Jansky was investigating the nature of radio noise, particularly that generated by thunderstorms that interfered with radio communication. In addition to noise of a terrestrial origin, he reported, "Radiations are received any time the antenna is directed towards some part of the Milky Way system, the greatest noise being obtained when the antenna points to the centre of the system." Radio astronomy was to prove of particular worth in understanding the violent nature of the cosmos. Until the advent of radio astronomy, the heavens were believed to be largely quiescent and unchanging. Radio astronomy was to change that reassuring picture.

The postwar reemergence of radio astronomy was led by a new breed of radio engineers, trained in the radar and radio direction-finding techniques of wartime, but so easily adapted to radio observations of the cosmos. The first discrete celestial radio source was identified in 1946 in the constellation Cygnus. By the late 1950s, catalogues were being produced of hundreds of objects in the radio sky, many of which could be identified with objects also detected in optical telescopes. Some radio sources were identified as lying within the Milky Way, and some were external galaxies.

Early in the history of radio astronomy, a very small fraction of galaxies were recognized to be particularly intense at radio wavelengths. They coincided with faint optical objects, and were given the name *quasars*. Although the optical objects looked almost star-like, it was demonstrated that they must in fact be galaxies at vast distances. Their radio brightness was such that the quasars had to be intrinsically at least 100 times brighter than any other known galaxy. A search of old photographic plates showed that quasars varied significantly in brightness over a period of just a few years. Nothing can travel faster than light. Hence, no object can coordinate its activity on its remote side with that on its near side in less time than it takes for light to travel across it. Thus, a change in intensity over, say, a 10-year period implies that such a quasar must be less than 10 light-years across, compared with, for example, the 100,000-light-year diameter of the Milky Way. So here was the dilemma: Not only were the quasars intrinsically brighter than any other galaxies, but also their energy was being radiated from a region a mere fraction of the size of a normal galaxy. Today, quasars are thought to be the highly active central nuclei of nascent galaxies at extreme distances. Early in the history of

SETI a famous incorrect identification of a quasar led to a false claim of detection of ETI by Soviet scientists (more on that little saga in the next chapter).

Today, massive radio telescopes, with dimensions measured in kilometers, provide astronomers with "radio spectacles" through which they can "view" the heavens in radio waves in considerable detail. The picture revealed by radio astronomy of a universe undergoing violent upheaval and change is dramatically different from the apparently quiet universe observed since antiquity with the unaided eye. Radio observations have made a major contribution to the current understanding of cosmic objects and phenomena. And radio telescopes now provide the powerful "ears" for SETI.

A conventional radio telescope usually consists of a large parabolic bowl to collect the radio waves and bring them to a focus, in the same way as does the concave mirror of an optical telescope. The larger the collecting bowl, the fainter the radio signals that can be collected (for example, a 300-foot-diameter radio telescope could detect the signal from a cell phone at a distance of some 300 million miles). Large radio telescopes are used for SETI. There is a limit to the size of the collecting bowl that can be constructed physically, especially if it is to have the capability to be steered to point to different positions in the sky. Giant telescopes with a diameter of approximately 300 feet that can be steered with precision have been built (the famous Lovell telescope at Jodrell Bank has a diameter of 250 feet). But to achieve larger collecting areas, various novel techniques are used. The 1,000-foot-diameter telescope at Arecibo, used for SETI, is slung in a natural ravine, and although it depends on the rotation of the Earth to sweep across the sky, some clever technology does give it a limited capability to direct the receiving beam.

The size of a radio telescope does not only determine the faintness of signal that can be collected; it also determines the telescope's ability to resolve the structure within a radio source or to distinguish two nearby radio sources. This is referred to as *resolving power*. Resolving power can be increased by sophisticated techniques combining the outputs of two or more radio telescopes. The emission from most celestial radio sources does not vary over reasonable time scales of days or years (pulsating sources called *pulsars* being a rather notable exception), and an array of small radio telescopes can be used in a technique called *aperture synthesis* to mimic the performance of a radio telescope of massive size, as the rotation of the Earth allows the array of small telescopes to sweep out a large aperture. (A development of this technique known as *very large baseline interferometry*, VLBI, combines signals received at radio telescopes separated by hundreds or thousands of miles.) The aperture synthesis approach is fine when the radio emission is constant in strength and when one can afford the time for the telescopes to sweep out the larger aperture, but would not be appropriate for SETI where a rapid response to a varying signal is needed. It is the big parabolic radio telescopes that have offered the greatest hope of snaring the elusive prize of an ETI signal.

Radio waves are a form of electromagnetic radiation. Other forms of electromagnetic radiation are visible light, X-rays, gamma rays, and infrared and ultraviolet radiation. These various forms of radiation were discovered independently, before it was realized that they were all manifestations of the same physical phenomena. All forms of radiation travel at the speed of light, and are characterized by their *wavelength,* the distance between adjacent wave crests or troughs. The number of wave oscillations passing a particular fixed point each second is called the *frequency*. Frequency is measured in cycles per second, which is also called *hertz*, abbreviated as Hz. All forms of radiation are emitted from stars and stellar systems, and can be detected with suitable telescopes.

Optical and radio telescopes can operate on the Earth's surface (although large optical telescopes are usually placed in the clearer air of high mountain tops). Telescopes to detect X-rays, gamma rays, and infrared and ultraviolet radiation are sent into space because these radiations do not penetrate the Earth's atmosphere (although a little infrared can). There are also benefits in putting optical telescopes into space, to free them from the annoying distortions to optical images caused by the atmosphere, but the high cost of space missions means that the vast majority of professional optical telescopes remain Earth-bound.

Infrared radiation comes from comparatively cool objects and systems, for example, dust clouds and planets, with temperatures measured in hundreds of degrees. Visible light is characteristic of objects with temperatures of thousands of degrees, such as the surfaces of stars, while ultraviolet radiation suggests temperatures of tens of thousands of degrees, characteristic of the outer atmospheres of stars. For X-rays to be generated, objects must be at temperatures of millions of degrees. An intriguing category of X-ray objects involves binary stars (two stars orbiting one another), where one of the pair is a dense compact object and the other is a normal star. Gas from the outer extremes of the normal star is drawn by gravity onto the compact object, like water flowing down the plug hole of a bath with the tap at the other end kept running. The material does not flow directly, but via a swirling disc (an *accretion disc*) where the gas is heated to the extreme temperatures needed to emit X-rays.

The basic argument for SETI initially put forward by Cocconi and Morrison has a theoretical component, a practical component, a philosophical component, and an appeal to the priority of experiment expected in legitimate science (the ultimate challenge for science is "prove it"). All of these are captured in the stirring final paragraph of their paper. The theoretical component is captured in the phrase "the presence of interstellar signals is entirely consistent with all that we now know." The practical component is covered by the phrase "if the signals are present the means of detecting them is now at hand." The authors turn to philosophy with the claim "few will deny the profound importance, practical and philosophical, which the detection of interstellar communications would have." And finally there is the appeal to the priority of experiment: "The probability of success is difficult to estimate; but if we never search, the chance of success is zero." These various components are all intended to support the final conclusion that "a discriminating search for signals deserves a considerable effort."

The beautifully written final paragraph of Cocconi and Morrison's paper set the challenge for the SETI pioneers—and still captures the spirit of the SETI programs presently being undertaken. As a clarion call for an area of scientific research, this final paragraph has had as profound an influence as any hundred words ever written in science.

There is a close relationship between the theoretical and practical components noted above. The real question is whether the existence of other radio-communicating civilizations in our own Galaxy is compatible with all we know. Actually, the most detailed current SETI project involves only the nearest 1,000 solar-type stars, this representing about one-hundred-thousandth of the Galaxy. Hence, what is really needed is a stronger theoretical commitment that ETI is indeed possible in such a relatively small population of stars. We need some indication that the mechanisms by which we believe intelligent life on Earth arose are sufficiently strong to have caused intelligent life also to be present, with a reasonable probability, somewhere in the one-hundred-thousandth or so of the Milky Way Galaxy that we hope to search in great detail in the foreseeable future. In the part of this book addressing the McCrea question, we will explore the basis for this stronger theoretical commitment.

The Green Bank Conference

A year prior to the publication of Cocconi and Morrison's paper, the young radio astronomer Frank Drake, working at the U.S. National Radio Astronomy Observatory in Green Bank, West Virginia, had already started planning just the sort of search proposed by the two Cornell physicists. By 1960, Drake's "Project Ozma" was under way. SETI was born into an uncertain world, of hope and vision more than matched by skepticism and funding uncertainties.

The simplest way to see the true function of the Drake equation in SETI is to consider the historical context at which it was first introduced, at an informal conference held at the National Radio Astronomy Observatory in Green Bank in November 1961. This conference was the first occasion on which the radio astronomers who were just beginning their search for extraterrestrial communications invited those from related disciplines to comment on their work. An interesting invitee was John C. Lily, who had just published his controversial book *Man and Dolphin*, in which he had claimed that dolphins were an intelligent language-using species. If anyone could comment on a plausible range of values for f_l and f_i, then surely Lily could. Another attendee, Melvin Calvin, a biochemist, was actually awarded during the course of the conference a Nobel Prize for his work on chemical pathways in photosynthesis. Bernard Oliver, an electronics expert from Hewlett-Packard and a remarkably influential presence in SETI history, was another key participant. Morrison attended, as did planet expert Carl Sagan, who was to become one of the great popularizers of astronomy, and of SETI in particular.

Trying to draw together scientists from diverse fields, Frank Drake, who was responsible for the scientific organization of the Green Bank conference, began by writing down the topics to be considered during the discussions. As he later recalled:

> I took on the job of setting an agenda for the meeting. There was no one else to do it. So I sat down and thought "What do we need to know about to discover life in space?" Then I began listing the relevant points as they occurred to me....
>
> I looked at my list, thinking to arrange it somehow, perhaps in the order of the relative importance of the topics. But each one seemed to carry just as much weight as another in assessing the likelihood of success for any future Project Ozma. Then it hit me: The topics were not only of equal importance, they were also utterly interdependent. Together they constituted a kind of formula for determining the number of advanced, communicative civilizations that existed in space.
>
> I quickly gave each topic a symbol, mathematician style, and found I could reduce the whole agenda for the meeting to a single line.
>
> Of course I didn't have real values for most of the factors. But I did have a compelling equation that summarized the topics to be discussed.
>
> My agenda equation later became known as the Drake Equation.... I'm always surprised to find it viewed as one of the great icons of science, because it didn't take any deep intellectual effort or insight on my part. But then as now, it

expressed a big idea in a form that a scientist, even a beginner,
could assimilate.

How a conference agenda piece became one of the great icons
of modern science is an intriguing bit of social history, reflecting
the interest people have in the prospect of detecting intelligence
elsewhere in the cosmos. Frank Drake became the "patron
saint" of SETI, and it is fitting that the equation that captures
the aspirations of SETI should carry his name.

Although the Drake equation has been at the heart of modern
SETI, it has not been without its critics. The critics claim that we
simply lack the theory to attach warranted values to most of the
terms in the Drake equation. The equation has been criticized as
falsely raising hopes of a detection of ETI, since it appears to
suggest that detailed scientific methodology is being brought to
bear to calculate a certain outcome. There are still critics aplenty, even after almost 40 years of the Drake equation being used
to argue for (and against) the reality (or otherwise) of ETI. The
Drake equation has survived all that the hostile critics have
thrown at it, and come through with its honor intact.

The Drake equation and SETI are based on the premise that
there is nothing special about planet Earth, that the conditions
on Earth that allowed intelligent life to evolve could be repeated
many times over elsewhere in the cosmos. Critics have attacked
even this premise.

In defense of SETI, it is clear that critics may have read more
into the Drake equation than was ever intended. The Drake equation simply does not play the role in SETI that is often claimed by
many critics, as well as some supporters. The equation was introduced (at the 1961 Green Bank conference) as a conceptual tool
for organizing discussion. Many fail to realize that what is presented as an equation need not function, fundamentally, as a
calculating device. It must be conceded that it is not only the critics of SETI who fall into this trap; many of the scientists actually
presenting arguments in support of SETI make a similar error.
However, if we are looking at the best practice of SETI proponents, then the criticisms are not valid. The Drake equation
serves to concentrate research on the issues that matter—and its

role in SETI therefore has been fundamental. What the Drake equation is really saying is, Here are the issues—here are the factors that we should really take account of—let's think about probabilities rather than certainties—let's search! And that seems to be the sort of challenge that science should be prepared to face.

There are several aspects of the equation that, while they may not have been of paramount interest to Drake in organizing discussion at the Green Bank conference, are worthy of some discussion. In writing the probability symbols in the sequence he did, Drake made heavy evolutionary associations. The equation has dead planets progressing to life, then progressing to intelligent life, and finally progressing to technological civilizations able to transmit radio signals. The manner in which these evolutionary factors are considered in the Drake equation differs substantially from anything in the common run of evolutionary theory; so it is hardly surprising that conventional evolutionists have found little to please them in the Drake approach. Certainly evolutionary biologists have not attempted to develop an analogue to the Drake equation to consider the factors that led to the emergence of life, intelligence, and then technology here on Earth. And if the evolutionary biologists have not been able to develop a product of probabilities equation for the evolution of life on Earth, where we know with certainty the outcome, it does cause them to ponder the validity of such an approach in an alien environment where the outcome is unknown. Some of the most severe critics of the Drake equation have come from evolutionary biology, although we would argue that many of their criticisms are based on a misunderstanding of the true purpose of the Drake equation.

In introducing the various probability factors, Drake was distancing himself from a vast tradition of arguments for the existence of ETI based solely on large numbers; that is, Why have all those stars in the heavens and no one to live there? Many critics of the equation fail to realize how Drake abandoned such simplistic considerations.

In writing the equation as the product of probabilities, Drake was setting up SETI as potentially vulnerable on many fronts. It is a crucial aspect of the relationship between the probability factors involved that if only one of them turned out to be vanishingly small, for example, planets were exceedingly rare, or the origin of life was an extremely improbable event, then the whole project was undermined. It is probably this property that has led to the Drake equation being as popular with SETI's opponents as it is with SETI's proponents. Opponents have tried to demonstrate that one or other factor must be vanishingly small, thus invalidating the whole of SETI. Some have argued that even the mathematical formulation is overly simplistic: "all those multiplications, and not a single addition—nature just isn't like that!"

Even at this point, the Drake equation is open to accusations of huge informational gaps, especially in the biological and social factors in the equation. While most astronomers recognize SETI as a legitimate endeavor, because they understand the nature of observations that can lead to satisfactory estimates of the astronomical factors, many scholars from other fields view the Drake equation with severe reservations.

Looking over these points, it is clear that the Drake equation does indeed embody a number of interesting commitments about ETI. These, rather than any sense in which it alone represents a strong computational tool that can justify the investment in SETI, are what have ensured its popularity. The history of SETI does little to undermine the interpretation that the main function of the Drake equation has been to focus investigation, as we will see in the next chapter.

A delightful summary of the equation was given by SETI pioneer Bernard Oliver, who described it as "a way of compressing a large amount of ignorance into a small space." This is a long way from the sense in which critics of SETI take it. The simple point that the Drake equation is a conceptual tool for organizing discussion and highlighting the areas in which the ETI hypothesis might be vulnerable counters many of the criticisms to which it has been subjected.

To support the idea that ETI will be sufficiently common to make SETI viable requires an argument that is best presented in two stages. The first stage involves the assumption that the solar-type stars that can now be observed should be relatively similar to our own Solar System at about the sort of stage when life started to develop on Earth, about 100 to 200 million years after the planet's surface cooled to the point where life could possibly become established. This is in many ways an amalgamation of Drake's factors f_p, n_e, and part of f_l; essentially we are asking what needs to be in place so that biology can start doing its bit.

The second stage of the argument then requires that, given physical situations similar to those of the early Solar System, there is an evolutionary pathway that increases significantly the probability of a biological situation arising similar to that of the present Earth. Here is where the evolutionary biologists start to get particularly anxious!

There are significant scientific questions relating to both parts of this argument. We believe there are billions of solar-type stars in the Galaxy, and a similar number of situations that may have been like the early solar system. Hence, the preferred biological pathway would have to be available to about one in at least a million such systems to give more than a few thousand planetary systems with intelligent life, even allowing for an optimistic estimate of other uncertainties. This is probably the hopeful extreme of our practical SETI capability in even the long term. We are susceptible to the illusion of the generously stocked shoe store!

We need to look beyond reason via large numbers, and remember that it is wrong to treat the Drake equation as if it were a means of calculating anything. The Drake equation's great value is in allowing scientists to martial their thoughts. To quote from the SETI Institute's Web home page:

> Within the limits of our existing technology, any practical search for distant intelligent life must necessarily be a search for some manifestation of a distant technology. A search for extraterrestrial radio signals has long been considered the most promising approach by the majority of the scientific community. Besides illuminating the factors involved in such a search, the Drake Equation is a simple, effective tool for stimulating intellectual curiosity about the universe around us, for helping us to understand that life as we know it is the end product of a natural, cosmic evolution, and for making us realise how much we are part of that universe. A key goal of the SETI

Institute is to further high quality research that will yield additional information related to any of the factors of this fascinating equation.

It is indeed a fascinating, and a beautiful equation. But beauty is in the eye of the beholder, and the beauty apparent to a majority of astronomers appears to be less evident to many evolutionary biologists and philosophers—and some politicians who have fought the funding of SETI!

Before progressing further with our consideration of the Drake equation, we need to consider the birth, life, and death of stars, since the first term in the Drake equation is all about the stars. Our life story of stars will be somewhat brief, and we apologize for the fact that it leaves many questions unanswered. However, we are exploring the evolution of life, rather than the evolution of stars, so that our consideration of many fascinating facets of astronomy must be cursory.

The Birth, Life, and Death of Stars

Stars are formed from isolated clouds of gas and dust in the interstellar medium. Such clouds of gas (known as *nebulae*) are observed illuminated by starlight, and form some of the most picturesque of astronomical objects.

As an interstellar cloud of gas (predominantly hydrogen and helium from the primordial universe) collapses under the effect of gravity, the energy of infall is converted into heat, so that the collapsing cloud soon attains an extremely high temperature, of the order 10 million degrees. At such extreme temperatures, certain *nuclear fusion* reactions can take place. In a newly formed star, hydrogen nuclei are "fused" together to form the heavier helium nuclei with the release of vast amounts of energy. The liberation of this *thermonuclear energy* increases the pressure in the mass of the gaseous material to the point where gravitational contraction is halted. A star is born.

The young star soon settles down to the relatively stable state in which it spends most of its active life. During this long period of stability, the star's self-gravity acting inward is balanced by the pressure pushing matter out. This delicate stellar balancing act is maintained at the expense of the loss of nuclear fuel. In a star like the Sun, about 655 million tons of hydrogen is transformed into about 650 million tons of helium each second. The lost mass is converted to the energy that eventually radiates from the star's surface.

And so it is with all the stars. The loss of mass and the generation of thermonuclear energy provide the answer to the question that has challenged human imagination over the millennia: What makes the Sun and stars shine? The secret energy source utilized with potentially catastrophic consequences by humans in the building of thermonuclear weapons (hydrogen bombs) is the same energy source harnessed and controlled in the central nuclear furnaces of the stars.

Although the nuclear fuel reserves of a star are enormous, they are not unlimited. When the hydrogen in the central core of a star is expended, gravity again takes control. As the core starts to contract again, it causes the internal temperature to increase to about 200 million degrees, when fusion of the helium ash left over from fusion of hydrogen takes place. Helium nuclei fuse to form carbon and oxygen; now the stars are generating the "stuff" of life! When all the helium in the core is expended, later stages of nuclear fusion may follow involving the fusion of successively heavier elements all the way to iron, beyond which no further fusion reaction can generate energy.

Thus, the long sought-after goal of the medieval alchemist, to change the elements from one form to another, has indeed been achieved on the cosmic scale in the centers of stars. But if the heavier elements such as carbon, nitrogen, oxygen, magnesium, sulfur, nickel, cobalt, and iron are formed in the inside of the stars, how are they released to the interstellar medium to contribute to the formation of new stars, planetary systems, simple life-forms, and ETI? This happens in a star's death throes.

The lifestyle and eventual fate of a star depend on its initial mass. Massive stars (10 to 100 times the mass of the Sun) shine the most brilliantly, but burn up their nuclear fuel reserves within a few tens of millions of years. On the other hand, stars of more modest size, like the Sun, will live for 10 billion years or longer. Massive stars may be unstable during periods of their evolution, and shed part of their outer fabric to the interstellar medium. For stars ten to twenty times the mass of the Sun, death comes in a spectacular blaze of glory. The accelerating consumption of nuclear fuel leads first to the expansion of the massive star to become what astronomers call a *red supergiant*. Eventually, with all the nuclear fuel reserves expended, gravity causes the central core of the star to collapse catastrophically, to form a rapidly spinning compact stellar remnant called a *neutron star* (where matter has been compressed to the extreme density of neutrons). A neutron star can be observed as a radio *pulsar;* a beam of radio emission sweeps around the sky like a lighthouse beacon as the neutron star rotates rapidly many times each second. For stars with very massive cores, gravitational collapse produces the ultimate state of compaction, a so-called *black hole,* with the gravitational field so intense that even light cannot escape it. The collapse of the core is accompanied by an explosive ejection of the star's outer envelope, witnessed as a *supernova* explosion. The energy released in these spectacular displays of celestial pyrotechnics is almost beyond comprehension; it is equivalent to the simultaneous explosion of 10 billion, billion, billion (10^{28}) 10-megaton hydrogen bombs! In the extreme conditions of these acts of stellar suicide, elements heavier than iron, such as platinum, silver, gold, and uranium, may be formed, enriching the interstellar medium with heavy elements. Such elements are rare, simply because the phenomena that create them, supernovae, are comparatively rare. The after-effects of a supernova explosion can be witnessed for hundreds of thousands of years, as an expanding *supernova remnant.* Some of the most spectacular nebulae in the sky, such as the famous Crab nebula, are the remnants of ancient supernovae. Supernovae are also believed to produce *cosmic rays.* Cosmic rays are energetic particles permeating the whole of space that can be detected bombarding the Earth's atmosphere.

The short life and likely catastrophic end of massive stars suggest that their planets are unlikely to bring forth ETI. Additionally, supernovae occurring relatively near to less massive stars with planetary systems could influence the evolution of

ETI, as we will ascertain later. In the extreme case of a very near-by star exploding as a supernova, ETI could be annihilated.

For stars of more modest size than those that evolve to become supernovae, when hydrogen reserves are expended, helium burning swells the star to become a *red giant*. However, gravity is insufficient in a small star to drive it to the later stages of nuclear fusion. When helium reserves are expended, the star contracts and cools to become what is known as a *white dwarf*. This peaceful form of death is believed to be the eventual destiny of 999 of every 1,000 stars.

The Sun is about 5 billion years old. In about 5 billion years its external hydrogen reserves will be expended, and it will start burning helium. Its increased energy will boil off the Earth's atmosphere and oceans. It will gradually swell to become a red giant, engulfing the planets Mercury, Venus, Earth, and Mars in its expansion, before gradually contracting into a geriatric white dwarf. Perhaps some advanced extraterrestrial civilizations, faced with the approaching oblivion of their parent stars, have had to venture out into the cosmos as an evolutionary imperative. Since the Earth at less than 5 billion years old is merely approaching middle age, the human race does not face such a pending dilemma. However, any life-forms on the Earth 5 billion years from now will be facing oblivion.

Factors in the Drake Equation

While avoiding the temptation to use the Drake equation merely as a calculating device, it is instructive to investigate the various factors included in it. To do this, we will convert it to its most familiar form:

$$N = R f_p n_e f_l f_i f_c L$$

Here we have replaced N^* by R, where R is a symbol representing the rate of formation of stars that might have planetary systems and that can survive long enough to be suitable for the development of intelligent life. (R is taken as the average rate of formation of suitable stars over the lifetime of the Galaxy.) Essentially what we have done is divide N* by the lifetime of the Galaxy to give the number of suitable stars that are formed each year. Having arbitrarily divided one factor by the lifetime of the Galaxy, we now have to multiply another factor by it to keep our equation balanced, and the factor most appropriately multiplied is f_L, the fraction of advanced civilizations existing at this time. This converts f_L to a factor L, the expected lifetime of any communicating civilizations.

Again, the message is do not be put off by a mathematical formula; we are still focusing the approach firmly on common sense. In all that follows, the use of the Drake equation should be accepted in this spirit of stimulating intellectual curiosity about the universe. Each of the Drake equation terms will now be investigated, as they help us to understand the issues behind the evolution of the cosmos and the emergence of life.

R
The rate of the formation of stars that might have planetary systems and that can survive long enough to be suitable for the development of intelligent life.

The Milky Way Galaxy is discus shaped, with a bulge in the center, almost like two poached eggs back-to-back. The stars within the Milky Way lie in distinct populations. The first of these, a sparsely populated spherical "halo" of stars surrounding the disc, are very old stars, likely to have been the first generation of stars that formed in the newly evolving Milky Way. Because they did not contain the heavy elements subsequently processed in successive generations of stars, they could not have formed with planetary systems, and there is little point looking to them for populated planetary systems. ETI will be found elsewhere. The second population of stars occupies a "thick disc" lying outside the main disc. Again, these are relatively old stars, typically twice the age of the Sun. Together the spherical halo and thick disc contain less than 10 percent of the stars in the Milky Way

Galaxy. The halo and thick disc are unlikely sites for ETI.

Most stars lie in the "thin disc" containing the spiral arms of the Galaxy. This structure of the Galaxy has been established primarily by studying radio emissions from the tenuous hydrogen gas lying between the stars. By sheer good fortune, neutral hydrogen radiates a characteristic form of radio wave.

Electromagnetic radiation can be spread out to form a spectrum of the component wavelengths by a device called a *spectrometer*. (A simple example of forming a spectrum is passing a beam of white light through a glass prism, to spread the white light into the colors of the rainbow—red, orange, yellow, green, blue, indigo, violet.) The nature of the spectra of electromagnetic radiation can tell astronomers a great deal about the composition of the emitting object, as well as of the tenuous material that the radiation has passed through along its journey through the cosmos to the telescope. The formation and interpretation of spectra, called *spectroscopy*, is one of the most powerful techniques of astronomy.

The characteristic radio waves emitted by hydrogen lying between the stars have a length of 21 centimeters. By studying the 21-centimeter radio emissions, radio astronomers have been able to map the shape of the Galaxy in its hydrogen gas, showing its intertwined spiral arms and central bulging region. The stars in the central region are hidden from view in visible light by the dust that permeates the plane of the Galaxy. But nearby stars can be observed, and are found to be aligned with the spiral arms.

In the solar neighborhood, there is no shortage of stars. And there is no shortage of the dense clouds of gas and dust that form the stellar nurseries for the formation of new stars. We can look deep inside such clouds, using the infrared radiation that escapes through the dust to be observed from Earth. What we see in these dense clouds are stars in the very process of being formed. However, astronomers studying the composition of stars (by spectroscopy) have concluded that the majority of stars in the solar neighborhood are not as rich in heavy elements as our own Solar System. It seems that by good fortune the Sun may have formed from an interstellar cloud that had been through many stages of stellar reprocessing. Thus, for our Solar System there was no shortage of heavy elements such as carbon, oxygen, silicon, and iron—just the "stuff" needed for planets and life-forms. There are certainly stars aplenty in the solar neighborhood of an age and heavy-element abundance comparable to the Sun's, but they are in a minority. ETI in the immediate solar neighborhood is likely to be rare indeed, because the majority of local star systems may not be sufficiently rich in the elements necessary for ETI to have evolved.

As we look toward the center of the Milky Way Galaxy, the situation seems to improve markedly for ETI proponents. Astronomers studying the chemical evolution of spiral galaxies find that the concentrations of the elements heavier than hydrogen and helium increase steadily from the outer reaches of any spiral galaxy toward its central bulge. The usual explanation put forward for this is that star formation in a spiral galaxy occurs earlier, and is subsequently more vigorous, toward its center, and the rate of star formation falls off toward the outer regions of the galaxy. Hence, in our search for stars that might have planetary systems and that might have survived long enough to be suitable for the development of intelligent life, things should start to look better as we move toward the center of the Galaxy. But there is a problem toward the center. The incidence of supernovae is expected to increase, which, as we will discuss later, could be bad news for ETI because a nearby supernova could blast ETI out of existence. It seems that there may be a doughnut-shaped preferred "ETI zone" in any galaxy where star formation is high enough, but the number of massive stars is not so great as to make the incidence of supernovae a major problem for ETI survival.

Stars are characterized according to their spectra, the classes being labeled O, B, A, F, G, K, M, R, N, and S. The alphabetical labeling in this form was something of a historical accident, but nevertheless astronomers have retained the classification scheme. It is easily remembered by the politically incorrect mnemonic: "Oh *be a f*ine *girl, k*iss *me r*ight *n*ow *s*weetheart!" (A version of the mnemonic more popular with female astronomers is "Oh *be a f*ine *girl, k*iss *me r*ight *n*ow." Smack!) Type O and B stars are blue, massive, bright, and short-lived, not the most likely parent stars for stable planetary systems bringing forward advanced life. They are called *early-type stars*. The R and later stars are red and small and have low intrinsic brightness. They are called *late-type stars*. While they are very long-lived, their low intrinsic brightness and smallness argue against their sustaining life-bearing planetary systems. The A, F, and G stars are yellow, with masses ranging from about half that of the Sun to about twice the Sun's mass. For each main class of star, there are ten subclasses. The Sun is a G-2–type star. It is among the A, F, and G (and possibly K and M) groups of stars that the prospects must look most encouraging for SETI—and humans stand as evidence of the suitability of these star types for advanced life.

R is the factor in the Drake equation that can probably be determined with greatest confidence. The observational evidence suggests a value for R of the order 20; that is, throughout the Milky Way Galaxy an average of twenty new stars of the type of interest to us form each year. One should not attach too great a degree of precision to this estimate. What astronomers are really saying is that a figure of 10 is probably too conservative and a figure of 50 would be too optimistic. A figure of 20 seems about right in terms of all we know, and for the purpose of SETI enables us to say with certainty that the rate of suitable star formation is certainly not a limiting factor. The Milky Way continues to churn out suitable stars, likely to have planetary systems rich in the "stuff" of life.

---

f_p
The fraction of those stars with planets.

The likelihood is that before a cloud of interstellar gas and dust collapses under the influence of gravity to form a new star system, it will have been rotating slowly (a gas and dust cloud is, in essence, a swirling eddy in a sedately rotating galaxy). As the cloud collapses, it will spin faster and faster, like a pirouetting ice skater pulling in his or her outstretched arms. There are two possible outcomes, depending on the shape of the cloud. In one scenario, the cloud breaks up into two separate blobs, forming two stars (a binary star system) rotating one about the other. (Very rarely three or more stars may be formed in a multiple star system.) The alternative scenario, of greater interest to us here, is that the collapse and accelerating rotation result in the formation of a flattened disc of gas and dust around the newly forming single star. Such flattened discs have been detected surrounding many stars in recent years, based on their infrared emissions. Within such flattened discs, local swirling eddies of dust eventually condense to rocks, which collide to become planetesimals, and eventually planetesimals collide to form planets. This model for the formation of a star and planetary system can explain why in our own Solar System the Sun rotates; why the planets orbit the Sun all in the same direction and in approximately the same plane, each with their individual "years"; why all the planets themselves rotate with their individual "days"; and why we see the residue left over from the formation of the planets in the form of orbiting asteroids and comets.

Only a few planets have been observed around nearby stars; however, the number detected is increasing steadily as new techniques are developed. While the resolution of the present generation of space-borne telescopes is not sufficient to actually see planets directly (other than in one unusual case), their presence can be inferred by various methods. The first method is to detect periodic shifts in the wavelength of the light from the parent star. The cause of this shift is a phenomenon known as the *Doppler effect*. The most familiar manifestation of the Doppler effect is in sound. When one hears a police car approaching with its siren blaring, the frequency of the siren sounds higher than when the car passes and recedes. At the point of closest approach, the frequency suddenly drops. Sound, like light, is a form of wave. When the police car approaches, the sound waves are bunched up ahead of the car, so that the sound is of a higher frequency than when the car passes and the sound waves are stretched out behind. So it is with light. When a source of light approaches, the light is shifted to shorter blue wavelengths, a so-called *blueshift*. When the light source recedes, the light is shifted to longer red wavelengths, a so-called *redshift*. The greater the speed of recession, the greater the shift toward red wavelengths.

(It is the Doppler effect that provided the observational evidence for the expansion of the universe, initiated by the big bang. The speed of recession of a galaxy is proportional to its distance, distance and speed being related by the so-called Hubble constant.)

The periodic variation of the wavelength of a star's light implies that the star is oscillating back and forth along the line of sight to the Earth. The simplest explanation for this is that the star is being orbited by another star (that is, we are looking at a binary system where only one of the stars is visible), or it is being orbited by a massive planet where the gravitational interaction between the star and planet causes the Doppler oscillation. Each option can be modeled, and in an increasing number of cases, the likelihood is that the observations are demonstrating the presence of a planet or planets. (In one famous case the star is a neutron star being orbited by at least three planets. This appears to be a system where the parent star exploded as a supernova, leaving a residual neutron star that retained at least part of the original star's planetary system, or perhaps the planets are fragments of a binary companion disrupted by the supernova explosion.) Alternatively, a "wobble" of a star may be detected back-and-forth across the line of sight. This requires particularly accurate measurements of the position of the parent star.

A different method for detecting planets is based on the fact that if a planet passes in front of its parent star, it blocks off some of the star's light. To determine the planet's orbital period (that is, its "year"), a second and third transit must be observed. It is believed that this technique could detect a distant planet even as small as the Earth, so long as the orbital geometry meant that the planet's transit across the face of its star lay along the line of sight to the Earth. A special space mission called *Kepler* is being planned to search for planets based on this technique. (Johannes Kepler in the seventeenth century put forward the first accurate description of how the planets follow elliptical orbits around our Sun.) Another major space mission, called *Darwin,* is being planned by the European Space Agency to search for planets around nearby stars and to detect evidence that they might be life bearing. The Darwin mission will have great value in trying to improve estimates of the next Drake factor, n_e.

Many of the planets being detected around stars are spectacularly *not* like the Earth, showing a wide variety of possible planetary configurations. The star 51 Pegasi is similar to the Sun in mass and temperature. But a planet recently detected in orbit around it has a short orbital period (that is, "year") of only 4 days. Therefore, it must be very close to its star. The closest planet to the Sun, Mercury, has a year of 66 days. Also, the Doppler velocity observations suggest that the 51 Pegasi planet must be approaching the mass of Jupiter, the largest planet in our Solar System. The implication is that the surface temperature of the planet is higher than about 800 degrees centigrade—certainly too hot for water-based life to survive. Several other Jupiter-sized planets have been found around other stars. This apparent preponderance of massive planets should not be misunderstood; they are merely the easiest to detect by the methods currently being used. An interesting recent result implies the presence of an Earth-sized planet orbiting a nearby star. A dust disc detected around the star has a missing doughnut-shaped ring, as if that part of the disc has condensed to an unseen Earth-sized planet.

There are still great uncertainties in determining the value of f_p, the fraction of suitable stars with planets, although these uncertainties will decrease markedly over the next few years if the rate of detection of planets continues to match the rate of recent discoveries. It is reasonable to assume that a value of 0.5 is not too far off the mark; that is, half the new stars created have planetary systems, while the other half probably evolve to binary star, or multiple star, systems. The value for f_p is unlikely to be higher than 0.5, but the recent rate of detection of planets (although still tiny in total) would suggest that a figure as low as 0.2 would present too pessimistic a picture. Until we have better data, 0.5 seems to be an entirely reasonable estimate.

---

n_e

The average number of planets per solar system whose environments are suitable for life.

Of the three astronomical factors in the Drake equation, n_e is the one that will long remain the most uncertain, and requires the most speculation in estimating its value.

Planets that can sustain life can be neither too hot nor too cold, neither too large nor too small; they must be "just right." To emphasize just how important temperature is to the pattern of life, if the Earth had been cooler to the point where chemical reaction rates (which are temperature dependent) were just a quarter of those we enjoy, then humans would not have evolved in the lifetime of the Sun.

Only planets larger than a certain size (and therefore with sufficient gravity) can retain an atmosphere (and therefore develop a biosphere). The Moon, for example, is too small to retain an atmosphere, even if the means to develop one had existed. A low gravitational field means that any gases emitted from a nascent planet would just drift off into the depths of space. But nor can a planet's gravity be too strong. Living things are made up of atoms and molecules, and the forces between them could not withstand excessive pressure. Thus, "normal" life could not exist on the surface of a neutron star; gravity would "squeeze" it out of existence. Similar restrictions apply with temperature. Since life as we know it requires water in which biochemical reactions can occur, if a planet is too cold, life will freeze; if it is too hot, life will cook. Any life-bearing planet must be "just right."

All stars of the type of interest (and which contribute to our estimate of R) have what is called a *habitable zone*. The habitable zone defines the range of orbits within which water and an atmosphere can exist. For the Sun, the habitable zone may initially have embraced Venus, Earth, Mars, and Jupiter (and its moons).

Humans can exist for a month without eating food, but will die within a week if they are deprived of water. Where there is no water, there is no life—at least life as we know it. It is hardly surprising, therefore, that water has held such symbolic importance throughout history. Water freezes as ice at 0 degrees centigrade and boils to produce steam at 100 degrees centigrade. Since life as we know it requires liquid water, it needs planetary conditions where, for at least part of the environment, the temperature range lies normally between the temperatures of ice and of steam. Earth fits the bill perfectly. In looking for ETI, we look for a habitable zone where liquid water can exist. Without precious water, there can be no precious life.

The surface temperature of the innermost planet Mercury is far too high to sustain life as we know it, even when the Sun was not as bright as it is now. Water and an atmosphere would have evaporated rapidly. Venus has now built up an atmosphere that traps incoming radiation (the so-called greenhouse effect, described in more detail later) so that the surface temperature is now far too high to sustain life, even recognizing that simple bacteria on Earth exist in hot springs and in the hydrothermal vents of the ocean floor. Mars certainly had substantial water reserves during earlier epochs, which are now preserved as permafrost at the poles. Simple bacteria might well have evolved on Mars, although as already noted, the evidence remains ambiguous. Jupiter is a gaseous planet. Its moon Europa has an icy surface, and its interior may be warm enough for the ice to have melted and simple organisms to have formed. The outer planets beyond Jupiter are outside the Sun's habitable zone.

Stars smaller than the Sun are cooler. Only planets close to such a star could lie within a habitable zone. But if they are too close, another problem arises. Planets very close to their parent star may be tidally locked so that they present the same face to the star at all times (just as the Moon does to the Earth). The atmosphere of such a planet would freeze on its remote side, and be unable to sustain life. A suitable planet also needs a magnetic field, so that its atmosphere is not blown away by the "wind" of particles emitted from the parent star.

In the case of massive, hot stars, the habitable zone may stretch a long way out from the star. But here the problem is the short life that massive stars have. As already noted, massive stars use up their nuclear fuel reserves very quickly and end their lives in spectacular supernova explosions. Even if primitive life-forms had formed in their habitable zones in the billion or so years of their existence, they would be blown to smithereens in the stellar holocaust.

So what would be a plausible value for n_e? Some have argued that a value of 1 would be acceptable, based on our own Solar

System. It is difficult to base any sensible estimate on just the one case we have information on. For our present purpose, it is probably best to acknowledge that n_e will long remain the most uncertain of the astronomical factors, and a value as low as 0.1 might be a more sensible assumption until data exist for a more optimistic estimate to be made.

---

f_l

The fraction of habitable planets on which life occurs.

Progressing to the biological terms in the Drake equation takes us from the astronomically plausible to the realms of evolutionary speculation. If the astronomers have been able to provide guidance on the earlier factors, albeit that this guidance is tempered with conditions, the evolutionary biologists have been unable (or unwilling) to proffer much guidance on this factor. The community itself remains divided between those who argue that life on Earth is a unique consequence of random events and those who might concede (but more in hope than based on any firm scientific evidence) that there are possibly preferred pathways for the evolution of life, and ETI may indeed evolve. There is no shortage of prebiotic materials in space. However, the likelihood that these could assemble randomly to produce the DNA molecules that lead to life is infinitesimally small. But DNA has appeared once, here on Earth, and the big question is, How? In Chapter 5 we will look in some detail at the origin of life.

There is little merit at this point to assign a range of values to f_l. We will return to this challenge at the end of this chapter.

---

f_i

The fraction of life-bearing planets on which intelligence evolves.

The *Encyclopaedia Britannica* notes that "it is difficult to imagine life evolving on another planet without progressing towards intelligence." Natural selection is based on survival of the fittest—or at least the survival of those best fitted to their changing environment. The type of catastrophic event of terrestrial or extraterrestrial origin that led to mass extinctions of life on Earth tens of millions of years ago (such as the demise of the dinosaurs) opened new niches into which surviving species were able to expand. (We will look at these mass extinctions in some greater detail later, in Chapter 4.) Given time, evolution appears to lead inextricably toward an enhanced physical capability and an emerging intelligence. Or does it? Evolution can get stuck in a rut. For billions of years, life on Earth remained in very simple forms. These simple life-forms existed over the eons in relative quiescence, avoiding predators and avoiding catastrophes that could have forced an alternative evolutionary path. What is it that forces nature to break out of such evolutionary dead ends? Of the billions of diverse life-forms on planet Earth, very few have developed intelligence.

It is difficult to be as optimistic as *Encyclopaedia Britannica,* which is essentially assigning a value of 1 to f_i. For our purposes here, let us take a value of 0.1. We need to admit that we really do not know, but compared with the inherent uncertainty in f_l, a measure of speculation in f_i will cause no harm in stressing yet again the difficulty of dealing with the biological factors in the Drake equation.

---

f_c
The fraction of planets with intelligence where radio communication potential develops.

Many species on Earth use very simple forms of communication. Humans, whales, and dolphins have developed "language"—not a very impressive sample from the billions of species that have inhabited the planet at some time, but nevertheless something to go on. The challenge in this factor is not communication per se, since the emergence of intelligence in the previous factor covers that. Here the challenge is the emergence of the technology to communicate by artificial means over a distance. However rapidly the intelligence of dolphins or whales evolves, they do not have the physical attributes to generate the technologies for radio communication while they remain limbless and in the sea. Intelligence alone is not the limiting issue in f_c; it is the motor ability to construct devices and systems which make up the technology that leads to radio or some other form of artificial communication. An elephant with the intellect of Einstein could not construct a radio telescope and receiver. If intelligence emerges, then f_c may be high, but it would be wildly optimistic to assign a value of unity. Again for our purposes at this stage, a value of 0.1 will do.

L

The lifetime of communicating civilizations.

And so we come to the final Drake factor. We have just one example on which to base our speculations—humans. The human race has had the technology to communicate by radio means for less than 100 years, and the technology for SETI for only four decades. For 50 years, it has also had the technology to destroy itself in a nuclear conflagration. As with f_l, we really have no way of knowing what a typical value for L might be—and we will leave the estimate unassigned at this point.

What Can We Make of All This?

While there is some prospect of improving our estimates of the astronomical factors, we must concede that for now f_l and L remain in the realms of mere speculation. Taking the most optimistic values for the other factors expressed, we can rewrite the Drake equation in a simplified form:

$$N = 10f_l L$$

With the more pessimistic estimates of the other factors, we derive the following:

$$N = 0.1f_lL$$

This is one-hundredth the optimistic value for N. This factor of 100 pales into insignificance against the uncertainties in f_l and L, so in the spirit of using the Drake equation as a basis of understanding the universe, it will suffice to write a simplified form:

$$N = f_lL$$

What does this tell us? If f_l is extremely small (say, one in a million), then L would need to be very large, say, 100 million years, in order for there to be even 100 potential communicating civilizations in the Milky Way Galaxy. Since the present-generation SETI instruments sample just one-hundred-thousandth of the Milky Way Galaxy, the chances of success must be deemed to be negligible if life is such a rare occurrence. It is only if f_l is very much greater than one in a thousand (probably at least one in a hundred) that there can be any prospect of a successful search, and only then if L is large.

What if advanced technological civilizations last for a relatively short period, say on average just 1,000 years? Such a very short period could result from "self-destruction" (for example, through nuclear conflagration, overexploitation of natural resources, famine and disease, or rampant pollution of a planet's atmosphere). Even if a civilization sorts out its own tendency to move toward self-destruction, natural perils await, for example, extraterrestrial phenomena of the type believed to have precipitated mass extinctions on Earth. Certainly a value for L as high as 100 million could be seen as optimistic (although not impossible), and a value as low as 1,000 years should not be dismissed.

If L was 1 million years, and f_l was very small (say one in a million), then N equals 1: us. If L was only 1,000 years, and f_l was an intermediate value (say one in a thousand), then N again equals 1: us. L and f_l both need to be "large-ish" to make the hunt for ETI worthwhile. We will address these issues again in Chapter 3.

We are back to the problem in the shoe store. Although we thought initially that we would be spoiled for choice, that turned out not to be so.

Observation

With all its uncertainties, has the Drake equation provided any insight to the quest for ETI? It is worth repeating, again, the words of wisdom from the SETI Institute Web home page:

> ...the Drake Equation is a simple, effective tool for stimulating intellectual curiosity about the universe around us, for helping us to understand that life as we know it is the end product of a natural, cosmic evolution, and for making us realise how much we are part of that universe.

In that spirit, no one could quibble with the contention that the Drake equation has provided an effective tool for stimulating intellectual curiosity—and it has helped us appreciate our place in the cosmos.

The story of our present cosmic quest is one that involves the most sophisticated instruments that technology is able to devise to extend the human senses. Great Earth-based telescopes and futuristic spacecraft continue to probe the depths of space to produce a multicolored panorama of the universe. They generate a sense of wonder and awe at the power of nature's forces and the beauty they create. Now we understand in considerable detail how the cosmos has brought forward stars, planets, and the basic building blocks for life-forms. But has it brought forward ETI? Everything science tells us suggests that it has. The prospects for SETI are tantalizing. Our attempts to understand the nature of the universe, and to seek out ETI, must surely represent one of the great human challenges, comparable to the exploits of explorers throughout the ages who sought to explore the unknown and to seek out new civilizations.

One of the true merits of SETI has been to encourage us to look at planet Earth as if from a distance and in general terms.

But once intellectual curiosity is stimulated, there is no escaping the need to carry out observations—to search, research, and research again. As Philip Morrison once noted, "It is fine to argue about N. After the argument, though, I think there remains one rock-hard truth: whatever the theories, there is no easy substitute for a real search out there, among the ray directions and the wavebands, down into the noise. We owe the issue more than mere theorising."

2. A Tale of Two SETIs

"The goal is not beyond us. It is within our grasp."
Frank Drake

It was the 1959 classic paper of Giuseppe Cocconi and Philip Morrison that set the theoretical and experimental basis for the pioneering radio SETI programs. The United States and Soviet Union each mounted SETI campaigns in the 1960s and 1970s, albeit approaching the challenge from somewhat different perspectives. This chapter explores the emergence of the U.S. and Soviet programs, how they diverged, and how common concepts were eventually established.

It needs to be remembered that the U.S. and Soviet programs were shaped in the 1960s and 1970s at the height of the cold war, and the two programs, with their differing approaches, became inextricably entwined with politics. The scientists involved seem to have largely stayed aloof from all this, but there is no doubting that Soviet politicians saw SETI as a research program where they could gain propaganda advantage over the United States.

Why Radio?

The Bishop of Birmingham, E. W. Barnes, proved to be an unexpected prewar visionary, when he wrote in 1931:

There remains the possibility of wireless communication...I

have no doubt that there are many other inhabited worlds, and that on some of them beings exist which are immeasurably beyond our mental level. We should be rash to deny that they can use radiation so penetrating as to convey messages to the Earth. Probably such messages now come. When they are first made intelligible a new era in the history of humanity will begin.

Any extraterrestrial civilization likely to go to all the trouble of setting up communication "beacons" would want them to be transmitting a signal that was likely to reach its intended destination. The main obstacles to this are the vast distances involved, and the tenuous gas and dust lying between the stars that can attenuate many forms of signal. Visible light can be scattered by interstellar dust; as beacons, radio waves have the attraction that they can travel through interstellar space with relative ease (although with some possible complicating factors, mentioned later).

The strength of any radio signal radiated uniformly into space decreases as the distance from its source increases, according to what is known as an *inverse square law*. If one doubles the distance from the source, then the strength of the received signal falls to a quarter; if the distance is doubled again, the signal strength falls to a sixteenth; and if it is doubled again, the signal strength falls to a sixty-fourth; and so on. Think of a garden sprinkler. The lawn in its immediate vicinity is well saturated while at a greater distance it is hardly damp. Eventually the lawn is outside the range of the sprinkler, and can only be watered by increasing the water pressure or moving the sprinkler. Thus it is with a radio beacon.

To transmit signals over vast distances requires very high radio power. Even then the signal received by a planet at a distance of, say, 8 light-years is only a sixty-fourth of that received by a planet at a distance of 1 light-year for a uniformly radiated signal. For interstellar communication, enormous power would be required for any transmitting beacon, and a very large receiving radio telescope would be needed to pick up the faint signals likely to be received over the vast distances of space. The most economic transmission technique for ETI would be to concentrate a radio beacon into a fine "pencil" transmitting beam, to sweep around the sky in a raster fashion, but this system would scan only part of the heavens on a given time scale. To cover a larger region of the sky in a given time interval, a beacon with a fan-shaped transmitting beam could be used to scan the heavens, although this would require much higher power than a pencil beam requires. To radiate a radio beacon uniformly in all directions out into the cosmos would require prodigious power levels. Thus, the most likely form of radio beam from ETI would be a pencil or fan beam occasionally sweeping past the Earth.

It is sometimes mistakenly thought that radio astronomers "listen" to signals, an understandable mistake for anyone who saw the heroine of the popular movie *Contact*, based around SETI, undertaking her task wearing headphones. But radio waves are *not* sound. It is true that radio pioneers figured out how to change sound into an electrical signal, which could be superimposed on a radio wave to be transmitted to distant receivers. There the signal was extracted and converted back into sound. Radio astronomy is about detecting electromagnetic radiation from the cosmos, *not* sound. Indeed, sound cannot travel through the vacuum of space (a fact that seemed to have escaped the movie makers of *Star Trek, Star Wars, Armageddon*, and the likes, who impressed us all with the roar of starships, death stars, and shuttles, in full quadraphonic glory). If you see radio astronomers wearing headphones, chances are they are just listening to their favorite music on their Walkman. The radio waves they are monitoring will appear as traces on oscilloscopes and data in their computers.

Magic Frequencies

The postwar years saw the emergence of radio astronomy as a powerful observational method, the new era being led by the radio and radar engineers who had pushed back the technological frontiers during the war effort. The sophisticated understanding that radio astronomers rapidly developed during the 1950s of the behavior of different signals enabled an informed choice of the appropriate signal for SETI. The best guess was a radio signal in the wide range of frequencies between 1,000 and 10,000 MHz. (MHz is the abbreviation for "megahertz," or 1 million cycles per second.) The lower frequency is limited by background radiation from the Galaxy that would swamp any desired signal, and the upper frequency is limited by the impact of atmospheric radiation that would produce a similar effect. Not surprisingly, a suitable instrument for looking for such signals was exactly the sort of radio telescope that radio astronomers were just beginning to make use of in the study of astrophysical signals in the late 1950s. However, this radio "window" from 1,000 to 10,000 MHz was far too broad to be scanned satisfactorily by the technology of the time, so some means was needed to narrow the range to search. Thus, Cocconi and Morrison introduced the notion of a "magic frequency" in their classic paper. They wrote:

> Just in the favoured radio region there lies a unique, objective standard of frequency, which must be known to every observer in the universe: the outstanding radio emission line at 1420 MHz (= 21cm) of neutral hydrogen. It is reasonable to expect that sensitive receivers for this frequency will be made at an early stage of the development of radio astronomy. That would be the expectation of the operators of the assumed source, and the present state of terrestrial instruments indeed justifies the expectation. Therefore we think it most promising to search in the neighbourhood of 1420 MHz.

This procedure of attempting to empathize with the ETI operators of a potential beacon proved extremely controversial, although the eventual emergence of a technology able to scan a very broad frequency range resolved the controversy in recent years. The problem was to decide which frequency should be chosen; there seemed to be an almost unimaginably large number of proposed magic frequencies. This made the whole question of trying to use such reasoning to narrow the possible range of frequencies distinctly problematic.

Later the complete range of frequencies of interest was narrowed to 1,000 to 3,000 MHz. This not only embraced the 1,420 MHz signal from hydrogen (H) but also the so-called 1,662 MHz hydroxyl (OH) line. It is the physical characteristics of the atoms and molecules making up any natural source of radiation that determine the nature of that radiation. Studying the spectrum of any natural source of radiation thus tells us interesting things about its composition. It was Morrison and Cocconi's contention that ETI would deliberately mimic the behavior of nature and transmit on a frequency, or frequencies, easily recognized by other intelligent civilizations.

Since hydrogen and hydroxyl make water, H_2O, the preferred radio range between the hydrogen and hydroxyl signals acquired the popular label of "the water hole." ("Where better to meet ETI than at the water hole" was the popular catch phrase.)

We need to define the term *bandwidth* of a signal at this point; it is the range of frequencies in which a desired signal is concentrated. Unwanted background interference is called *noise*.

Nature may be quiet in the water hole, but it certainly is not absolutely silent. However, natural sources of emission in the water hole tend to be broad band (that is, they embrace a wide range of frequencies) rather than the narrow band signals that it was believed ETI would certainly use. Narrow bandwidth signals ensure a high signal-to-noise ratio (the desired signal does not have to compete with a cacophony of background interference that would be present in a broad frequency band, so a lower level of transmitted power would be required).

Yes, ETI would have worked out the obvious benefits of using a narrow band signal in the water hole—or so it was passionately argued. Well, SETI had to start somewhere, and in attempting to read the mind of ETI, Cocconi and Morrison stimulated the theoretical considerations that would shape SETI.

Although radio frequencies were given prominence by Cocconi and Morrison's paper, an intriguing alternative was proposed by U.S. physicist Freeman Dyson a short time later. Dyson based his proposal on the assumption that there are civilizations millions of years more advanced than our own. Dyson suggested that such an advanced civilization could have developed the impressive technology to redistribute the material from one or more of its parent star's planets into a sphere of space colonies that orbit and absorb energy from their parent star, thus maximizing the utilization of the star's energy resources. This "sphere" of orbiting colonies (subsequently called a *Dyson sphere*) might then be detected by Earth-based telescopes from its substantial infrared emission from waste heat. Thus, such advanced civilizations could be detected by their impact on their space environment, rather than from any deliberate attempt they might make to communicate via a radio beacon. In Chapter 7, we will return to this possibility.

Other indirect measures of civilization were proposed. It was suggested, for example, that the detection of the 1,516-MHz radio emission line from tritium (a form of heavy hydrogen) near a star could indicate leakage from an ETI civilization's orbital nuclear fusion reactors. Equally imaginative were suggestions that ETI could dump waste from its nuclear reactors into its parent Sun, evidence for which could be seen in the star's spectrum.

Indirect signs of life could be sought in the atmospheres of any planets detected around distant stars (although such measurements are not yet possible with current instrumentation). For example, the detection of highly reactive gases in the spectrum of a planet's atmosphere (such as oxygen or methane) would suggest a biological source replenishing the atmosphere. Of course, this biological source might merely be "slime"—something more interesting would be required to prove that the planet was inhabited by ETI.

Cocconi and Morrison's paper unleashed a flood of imaginative proposals. But at this time radio searches remained the most persuasive, since it was argued that ETI would surely select a communication method that was easy and economical to generate and receive; would go where it was sent with the minimum of absorption or deflection; and would travel at the maximum speed known, the speed of light. Radio waves seemed to fit the bill perfectly. So radio was where SETI started, although in later years serious consideration would be given to SETI in other radiations.

Project Ozma

The man who would turn Cocconi and Morrison's proposals into reality was Frank Drake. Drake was a graduate of Cornell. He had long daydreamed about the prospects of communicating with ETI. He got the chance to turn daydreaming into real observing when he joined the staff of the newly formed National Radio Astronomy Observatory (NRAO) at Green Bank in 1958 (the year before Cocconi and Morrison's paper was published in *Nature*). The NRAO was funded to construct 140-foot- and 300-foot-diameter radio telescopes; however, to get into business quickly the NRAO staff first constructed one of more modest size. An 85-foot-diameter telescope was completed in April 1959, and Drake had persuaded colleagues that he should be allowed to use it for SETI. His chances of getting such approval at a more established observatory would have been low indeed. But in 1959, NRAO was new, its staff included young enthusiasts like Frank Drake eager to try unconventional things, and the acting director (Lloyd Berkner) was a research entrepreneur. When the first NRAO director, Otto Struve, was appointed in July 1959, it was Drake's good fortune that Struve had long been a believer in life on other worlds. SETI was about to become a reality.

Drake calculated that the 85-foot-diameter radio telescope could detect any ETI signals from stars to a distance of about 10 light-years, if transmitted with strengths comparable to intense radar signals from Earth. That was not very far but it would do for starters. A new ultrasensitive 1,420-MHz radio receiver was designed, and could be used for conventional radio astronomy projects as well as Drake's search. It had a signal band and a nearby comparison band, the principle being that taking the difference of these two signals would reject any spurious background signal, leaving only the desired narrow band signal from ETI. The receiving frequency could be tuned around the 1,420-MHz central signal, to allow for the suspected Doppler shift of ETI's planet around its parent star.

Drake called his planned search *Project Ozma*, after the movie *The Wizard of Oz* with its exotic "alien-like" characters. It was a low-budget project, just $2,000 being assigned to it. Even in 1960 that was not much cash for a complex research project. And it was to be kept secret. Struve was worried that as a new research establishment, NRAO could be criticized for wasting time and money on what was a highly speculative and somewhat unusual venture. There was concern about the "giggle factor," which could inflict any research project appearing to be too wacky to fellow scientists (and SETI certainly did suffer from the giggle factor in abundance during its early days). However, this well-intentioned secrecy about Ozma meant that the SETI limelight was stolen by Cocconi and Morrison's paper. Such was the scientific excitement generated by Cocconi and Morrison's paper that Struve much regretted his desire for secrecy, and just 2 months after its publication, Struve announced Project Ozma at a lecture he gave at MIT. Now NRAO could receive some credit for their well-advanced plans. Press interest was intense; Ozma was in the media spotlight. But the research community was divided. Those excited by the prospects of Project Ozma were countered by respected scientists who criticized it as an outrageous waste of money (despite its modest budget) and as an inappropriate use of valuable telescope time. Born in such controversy, SETI divided the scientific community for more than two decades. Its present comparative respectability (at least with a majority of scientists) was hard won. The giggle factor was not easily silenced, however, and even now reemerges in certain quarters on occasions.

Ozma observations commenced on April 8, 1960. The target stars were Tau Ceti and Epsilon Eridani, the two nearest solar-type stars known. Drake and his colleagues completed 200 hours of observation, searching either side of the 1,420-MHz magic frequency, to allow for planetary orbital motion. In fact, 7,200 channels of 100-Hz bandwidth were scanned on each target star. The result? Nothing! It just was not going to be that easy. The full extent of the radio spectrum where the background noise in the universe is quietest is some 100 million channels of 100-Hz bandwidth, and there are billions of possible target stars; hence, the 7,200 channels scanned by Ozma, and the targeting of just two stars, did not really amount to very much. Nevertheless, Ozma had pushed open the SETI door. Drake realized that receivers with millions of channels would be needed before SETI could get really serious, and for many years, until new technology was available, he moved into other areas of research (notably radio observations of planets and pulsars). However, for SETI there was no turning back, and Drake bided his time until the appropriate technology became available.

The imagination of creative thinkers had now been stimulated. Critics there would be aplenty. But just as science could never be the same once Copernicus had challenged Earth-centric thinking, so the universe could never be the same once the SETI pioneers had challenged human-centric thinking.

Drake's Ozma methodology was surprisingly similar to that proposed by Cocconi and Morrison, although developed entirely independently. The contact between Drake and the theoretical work of Cocconi and Morrison occurred at the Green Bank conference described in Chapter 1, where the Drake equation was first introduced. The consensus established at this conference remained remarkably stable, with very few shifts in emphasis, as the doctrinal hard core of SETI in the United States for the next three decades. The key elements of this doctrine can be summarized as being that the U.S. SETI community believes the following:

1. What is being searched for is deliberate attempts at communication by intelligent civilizations. Often a misleading impression is given that what is being searched for is the background radio signals associated with a civilization, such as its television broadcasts. This is not the case. What SETI is searching for are beacons deliberately announcing the presence of ETI.

2. These deliberate attempts at communication are likely to occur in a certain narrow radio band, since it is only such signals that would have much hope of traversing interstellar distances. These signals are to be searched for using large radio telescopes.

3. Although selecting the radio band enormously narrows the

range to be searched, too many frequencies remain to make searching them all practical. Hence, SETI searchers were licensed to make guesses, based on reasoning such as Cocconi and Morrison's, of magic frequencies where communication is most likely. The water hole containing the hydrogen and hydroxyl emissions has found particular favor.

4. Searches should be targeted at nearby Sun-like stars, since it is a reasonable guess than any communicating civilization will be found around such stars and our detection capabilities are decidedly limited in range. However, there might be some limited scope for all-sky surveys for high-power transmissions from distant ETI civilizations with a particularly high level of technological capability.

5. Since nearby stars constitute only a tiny fraction of the stars in the Galaxy, let alone the universe, there is a requirement that either we are extremely fortunate (there happens to be a needle in our chosen haystack) or that ETI must be comparatively abundant. It is this requirement of abundance that is the main target of many of the critics of SETI.

6. ETI is likely to be significantly more advanced than we are. Our Sun is not an old star. On average, stars are several billion years older than the Sun. Therefore, we should expect ETI to exist and be developed far beyond our present intellectual and technological states.

This set of assumptions and commitments, although not formally recorded in the manner we have listed them here, has remained more or less stable throughout the history of U.S. SETI. The notable exception is that magic frequencies have largely gone out of fashion. This occurred for a number of reasons. First, no signals have been found on the magic frequencies that have been explored. Second, it has been generally accepted that magic frequency reasoning does involve a degree of assuming human-like behavior for the signaler that may be unacceptable. We simply cannot guess what subjects ETI scientists might find most interesting. Last, and probably most important, tremendous advances in electronics enable the whole of a wide range of radio frequencies to be examined simultaneously so that the selection of discrete magic frequencies has been rendered largely redundant.

Soviet SETI

The initial development of SETI in the Soviet Union directly followed the work of Cocconi, Morrison, and Drake. Radio astronomer Iosif S. Shklovskii was inspired by this work to write *Vselennia, Zhivin, Razum* (Universe, Life, Mind), published in 1962 and the first book by an astronomer to give prominence to the radio search for ETI. Yet Soviet interest was inevitably colored by the distinctive philosophy required by the state. As Soviet astronomer Nicholas Bobronikoff wrote in his essay "Soviet Attitudes Concerning the Existence of Life in Space":

> The Soviets are emphatic that their materialistic philosophy is in complete agreement with the idea of extraterrestrial civilisations. According to this philosophy life is a normal and inevitable consequence of the development of matter, and intelligence is a normal consequence of the existence of life. Even the best informed scientists in the USSR, like Oparin and Shklovshiy, must necessarily subscribe to this crude philosophy promulgated more than 100 years ago by Marx and Engels.

Most of the commitments involved in Soviet materialism were actually shared with Western scientific materialism, although rhetorical flourishes often obscured this. It was, however, Marxist historical materialism's commitment to the inevitable technological development of civilization that gave Soviet SETI its most distinctive characteristics.

These characteristics were developed in the work of Shklovskii's former student, Nicolai Kardashev, at the Sternberg Institute in Moscow. He developed a framework for SETI that showed far greater imagination than the Americans'. In his 1964 work "Transmission of Information by Extraterrestrial Civilisations," he proposed a threefold classification of civilizations. It was suggested that type I civilizations would have a technology level whereby they mastered approximately the energy from their own star incident on their planet; Earth was intended to fall into this category. Type II civilizations would be capable of harnessing all the energy of their star. This would clearly require engineering on a quite stunning scale. (Dyson spheres come into this category.) Type III civilizations would far outstrip even these proposals and utilize energy on the scale of their own galaxies. (It was not even speculated how this might be done). Kardashev believed that the search for type I civilizations, as effectively emphasized in the U.S. radio search strategy, was futile, very much a case of chasing a very small needle in a gigantic cosmic haystack. In his view, it was more practical to try to hook the bigger fish of type II and III civilizations, which were supposed to be detectable whether or not they were signaling to us intentionally, and would be accessible even out to vast distances (indeed within other galaxies). It was argued that the universe had had plenty of time to produce ETI civilizations hundreds of millions of years older than ours, and with an intelligence and advanced technologies that it is impossible for the human mind even to conceive of. The type II and III civilizations were sure to make their presence known, so why waste time, the Soviets argued, looking for civilizations with technologies little more advanced than our own. Think big, think the future, think the unthinkable—that was the Soviet approach. Soviet SETI thinking was based firmly on the notion of *superiority*: the supposed superiority of Soviet SETI over U.S. SETI, and the imagined superiority of ETI technology over that of humans.

At the Soviet equivalent of the Green Bank conference, which took place in 1964 at the Byurakan Astrophysical Observatory in Soviet Armenia, Kardashev's proposals proved extremely influential. Hence, the aim of the conference was "to obtain rational technical and linguistic solutions for the problem of communication with extraterrestrial civilisations that were very much more advanced than our Earth civilisation." This approach of concentrating on highly advanced type II and III civilizations continued throughout the 1960s and 1970s, when Soviet SETI was gaining momentum, while in the United States scant thought was being given to post-Ozma surveys.

Although they did not discuss alternatives like interstellar travel and probes, as some scientists in the United States had done, and did not use a Drake equation approach, the Soviets addressed many of the same issues as the Green Bank delegates had. They paid more attention to the likelihood that extraterrestrial civilizations would be much more advanced than ours, and emphasized communication with ETI rather than merely searching for signals. They gave more attention to the problems of message decoding and problems related to the development of civilizations. Thus, there were distinct differences between the Soviet and the American approaches. Unlike the U.S. researchers, the Soviets were searching for incidental signals, rather than making deliberate attempts at communication. This resulted in them engaging in surveys of areas of the sky, rather than the favored American methodology of targeted search on local solar-type stars. The Soviets were confident that communication would eventually be established. The need to understand a basis for communication often led them to try to imagine the way ETI would think, much as Western researchers did while trying to calculate magic frequencies. The Soviets were looking for civilizations vastly more advanced than our own (the types II and III), while the Americans were searching for civilizations of comparable or slightly greater advancement (type I civilizations). Thus, while the Americans required an assumption of abundance of extraterrestrial civilizations with at least a modest radio transmission capability, the Soviets required far fewer with very advanced technical capability. They were dependent on an assumption of advancement: that civilizations could reach the prodigious heights described by Kardashev. This assumption of advancement clearly connects to historical materialist orthodoxy. Equally, the shift from abundance to advancement explains why Soviet researchers were not nearly so interested in the Drake equation as were U.S. scientists.

Contrasting the Soviet and American approaches in light of their situation in a world feeling the chill of the cold war shows clearly the influence that extrascientific factors can have on scientific practice. It is particularly ironic that these difficulties were manifested in SETI, since this was one of the few fields of research that saw a great deal of interaction between scientists from the East and West even at the height of the cold war. Although those involved, so excited at any contact at all, seem rarely to have realized it, they were divided theoretically in ways that reflected cold war ideology. To put it in somewhat simplistic terms, while Americans searched for "Radio-free Alpha Centauri," the Soviets sought to join the future meetings of the "Communist Party Intergalactic."

The U.S. and Soviet SETI research communities at the time did not really believe themselves united, but they did believe that they were divided purely because of the rivalry between their governments. When they were allowed to meet, they treated each other with great respect (Shklovskii and Sagan formed a particularly close partnership), and from a purely scientific perspective they would have understood themselves as more closely united than they were. On occasion, though, the genuine differences between them did seem to lurk in the background of discussion. However, history has made this criticism of the practices involved irrelevant, partly through the collapse of the Iron Curtain, and partly because the majority of astronomers now favor combining to some extent the sky surveys used by the Soviets and the targeted searches favored in the United States during the early years. The Soviet approach did demonstrate an inadequacy of the Drake equation, in that it has no factors to consider specifically advancement beyond the type I stage to the more advanced type II and III stages where detection of artifacts of their exceptional energy consumption would be possible.

The early Soviet program produced one of the great SETI false alarms. Sholomitskii, Kardashev's boss at the Sternberg Institute, had observed various radio sources between August 1964 and February 1965. One of these, a radio object called CTA 102 (an identification number in a catalogue of radio sources), was found to vary in strength by about 30 percent every 100 days. Influenced by Kardaskev's thinking about the abundant energy utilization of the proposed type II and III civilizations, and believing that CTA

102 lay within the Milky Way, Sholomitskii reasoned that he was observing powerful radio transmissions from another civilization. On April 12, 1965, he released a press notice through the state information agency TASS, suggesting that CTA 102 represented signals from a highly advanced extraterrestrial civilization. A high-level press conference was held in Moscow a couple of days later. But unbeknownst to the Soviets, the U.S. astronomer Maarten Schmidt had shown unambiguously that CTA 102 was a perfectly ordinary quasar. The Soviets were left looking somewhat foolish with their possible ETI "discovery" announcement. But lessons were drawn from this rather farcical incident, and today the international SETI community has accepted that any detection should require independent verification before any public announcement is made.

The first international SETI conference was held in 1971, 10 years after the Green Bank conference, at the Byurakan Astrophysical Observatory in the Soviet Union. Sagan, Drake, and Morrison were the U.S. organizers, and Shklovskii and Kardashev were the organizers from the USSR. International conferences are an important means of allowing scientists to announce their results and exchange ideas. Being invited to present a "keynote" lecture to an international conference is a form of recognition much prized by researchers. (The fact that international conferences tend to be held in sun-drenched climes in the summer, or leading ski resorts in the winter, is an added bonus for scientists. Many a productive scientific collaboration has been sealed on the ski slopes, or a new theory expounded in the surf!) International conferences shaped global cooperation in SETI to an impressive extent, even across the Iron Curtain at the height of the cold war.

The Byurakan conference confirmed for U.S. participants the massive scale of the Soviet effort. In their 1960s SETI program, Soviet politicians had seen the opportunity to emphasize their perceived technical lead over the United States in the post-Sputnik era. The U.S. Apollo program, however, more than addressed the technical balance. At Byurakan the two key SETI nations were able to confirm each other's optimistic prognoses for successful SETI. Drake recognized the validity of the

Kardashev analysis, and presented an alternative search strategy based not merely on the closest stars, but on regions of the sky where stars were most dense (for example, toward the center of the Milky Way, a prime search area for the reasons given earlier). He recognized that the intrinsically brightest radio beacons could indicate rare but highly advanced civilizations, and such beacons at greater distances could be easier to detect than faint emissions from nearby beacons. Although the U.S. approach would remain centered on the Drake equation, some aspects of the Kardashev strategy were now gaining recognition in U.S. planning.

Even while the Byurakan conference was in progress, NASA was publishing an optimistic assessment of SETI prepared by an Ozma veteran, Bernard Oliver (of Hewlett-Packard), and a NASA scientist, John Billingham. The report proposed a major new facility called Project Cyclops, a massive circular array of 1,500 radio telescopes to be dedicated to SETI. This represented an engineering challenge of Apollo proportions. The year 1971 represented the peak of optimism in the future of SETI based on the Drake equation. However, powerful critics were starting to emerge. Things would never be easy again for the SETI enthusiasts. Inadvertently, the Cyclops report had fed the SETI critics some powerful ammunition. The price tag looked staggeringly high, and there was scant recognition given by the critics to the fact that Cyclops could have been built up gradually over time. It certainly did not need the full 1,500 radio telescopes from the outset to make it a powerful SETI tool. The grand Cyclops vision did not get off the funding starting blocks, because SETI critics saw it as the ultimate scientific extravagance. The critics presented SETI scientists as being detached from the realities of the funding of research. They appeared to their opponents to be like spoiled children, demanding new and bigger toys.

Shklovskii was an unexpected defaulter from the Soviet SETI cause. It appears that he grew very skeptical about the possibility of any civilization surviving for long with the technology that not only makes possible interstellar communication but also allows mass destruction. The cold war between the United States and the Soviet Union left him so depressed, it seems, that he felt no ETI could survive their own "cold wars" for long. In Drake equation terms, Shklovskii decided that L would be on average sufficiently small that there would be only a very remote prospect of civilizations progressing to type II and type III stages. In Shklovskii's mind, technology would prove to be the path to self-destruction. SETI, sadly, lost one of its most enlightened supporters and greatest scientists to cold war depression. Shklovskii was a brilliant man, but as a Jew he never received the scientific recognition he deserved from the Soviet system. However, in the West his true genius was acclaimed. He had been a great friend of SETI, at a time when great friends were sorely needed. SETI would miss him.

Post-Ozma

Following Ozma, SETI in the 1960s was left largely to the Soviets. Despite the initial excitement over Project Cyclops, the emergence of powerful critics to SETI killed any hope that the high levels of funding required would be forthcoming. In 1971 Gerritt Verschuur repeated an Ozma-type search at NRAO using the 140-foot- and 300-foot-diameter telescopes. He called his survey Oz*pa*, to establish gender balance with Ozma! He spent just 13 hours searching nine stars, but this was with a receiver handling 384 frequency channels at once (in contrast to Ozma's single channel). Again, nothing. Ozpa was followed by Ozma II, undertaken by Ben Zuckerman from the University of Maryland and Patrick Palmer from the University of Chicago. They used the 300-foot NRAO telescope, and the same 384-channel receiver as Ozpa, to survey 674 stars between 1972 and 1976. Yet again, nothing. (In later years, Zuckerman was to become a SETI skeptic. It is not obvious what brought about his change of heart. It certainly should not have been the negative outcome of his Ozma II survey, since these pioneering searches were always going to be highly speculative. They were "pathfinders," identifying the best way forward for those who would follow, but with little prospect of completing the journey themselves.) Astronomers in Australia, Canada, and France also chanced their arm. But still nothing. There was certainly no denying the enthusiasm, professionalism, or sincerity of the pioneering searchers. But what was still lacking was the technological muscle, the carefully thought-out observing strategies, high-performance computers, and the large amounts of dedicated observing time needed for any real chance of success. The approach on these early surveys was "let's grab what telescope time we can and let's use the equipment that is readily available and let's take a guess at which nearby stars to point the telescope at." It was not going to be that easy to snare ETI. The real needs were long periods of dedicated telescope time, receivers customized to the special requirements of SETI, and cash.

Actually, even one dedicated radio telescope was not going to be enough for serious SETI. Two radio telescopes separated by some distance are often used. This is to help distinguish between interference from terrestrial sources, such as military radars, and signals that might be extraterrestrial. The effect of Doppler shift due to the Earth's rotation means there will be a slight difference in frequency in the signal received at the two telescopes if it is genuinely of extraterrestrial origin.

U.S. SETI's early need for dedicated telescope time was met from an unlikely source. An electrical engineer named Robert Dixon, from Ohio State University, worked with one of the pioneers of radio astronomy, John Kraus. Kraus had designed a novel radio telescope, affectionately known as "Big Ear." Big Ear was not pretty, but being the size of three football fields, it was certainly effective, and it produced one of the pioneering surveys of the radio sky. In 1972 funding for the Ohio radio survey ran out, and all the staff employed on the survey were laid off. Fortunately, they all found alternative employment.

Dixon, whose interest in ETI had been sharpened by working on the Cyclops study, suggested to Kraus that with the loss of survey funding, the Big Ear should be dedicated to SETI. Kraus agreed. And so started the world's longest-running SETI survey. The survey was finally abandoned in the summer of 1998, when the expiry of the university's lease led to the telescope's land being purchased for a golf course. The bulldozers moved in. It was a sad ending for a noble pioneer.

The Ohio SETI survey scanned the whole sky, rather than making a dedicated search of selected stars. Any unusual detections were then followed up. Big Ear was always short on funding, and the Ohio SETI program was run by volunteers, many of them being those previously employed on the radio survey program.

Big Ear detected one of the most tantalizing signals from any SETI program. On August 15, 1977, Ohio volunteer Jerry Ehman was scanning the computer printouts in the usual way, looking for anything that might represent a signal worth following up. Like the other Big Ear volunteers, Ehman had a full-time job elsewhere, and SETI was a hobby for evenings and weekends. For the volunteers at this time the visual scans of the data were a tiresome and largely fruitless task; present SETI surveys have the luxury of computers scanning the records in real time, to sound an alert if anything interesting is found. To Ehman's amazement, he saw a signal so strong that it had driven the Big Ear detector right off the scale. In his excitement, Ehman scribbled "WOW!" on the printout. To this day the detection has been called the *Wow signal*.

With any such signal, it is important to consider the options. Was it terrestrial interference? Could it have been a transmission from an orbiting satellite or an interplanetary probe? Was it a military radar, or an aircraft flying overhead? None of these or other possible explanations fitted the timing and nature of Wow. Yet numerous attempts to redetect the signal failed. The Wow signal remains one of the great mysteries of SETI. Was it an ETI beam momentarily sweeping across the Earth. Possible—yes; certain— no. In science, it is best to stick with the certainties rather than the possibilities. The Wow signal was real enough, and there was no plausible terrestrial origin for it. But without a confirmatory redetection, it could not pass the essential scientific test of validation. Hence, it remains one of SETI's mysteries, rather than the first positive ETI detection (which it plausibly might have been).

The inability to follow up the Wow signal stressed the desirability of having an automatic detection capability for possible ETI signals, so that observers could be alerted immediately and appropriate action taken. Computers were needed if real progress was to be possible.

For SETI, computers are now used to control the radio telescopes to make sure they are pointing to the right parts of the sky with the required precision at the expected times, to ensure that there is proper communication between remote telescopes, to monitor the correct operation of the receivers, and to analyze the received radio waves to look for possible evidence of ETI. Various checks have to be made to test the validity of any ETI alert against agreed criteria; there is no point to sounding the alarm if the signal is clearly terrestrial interference. Sophisticated analysis programs have now been developed to filter out any spurious detections. And all this has to be done at great speed, analyzing the outputs of the millions of channels now sampled simultaneously by modern SETI. Just to emphasize the computing power now engaged in SETI, the Project Phoenix MultiChannel Spectrum Analyser (more on this later) uses 384 powerful computer chips for processing the data from the radio telescope. The system can perform 75 billion calculations each second! Such are the computing demands of modern SETI, and the number-crunching power of modern computers.

A Billion Channels Please

U.S. SETI has been based on the premise that any narrow band signal is highly likely to be from an intelligent source. Certainly concentrating transmitter power into a narrow bandwidth is the most effective way of sending a radio signal over a vast distance, and dealing with background noise (from spurious sources of natural radiation). Everything in our scientific experience supports the premise that ETI would use a narrow bandwidth for targeted signaling, although if one was merely looking for background radiations from advanced technological type II or III civilizations (as the Soviets were), then a broader radio bandwidth could be used.

Naturally occurring radio emissions from the cosmos, for example, from quasars, pulsars, or supernova remnants, occupy a range of frequencies broader than about 300 Hz (usually very much broader). A signal concentrated into a narrow bandwidth of just a few hertz or less is surely going to be from an intelligent source. Think about sweeping across the dial of your radio set, seeking out a favorite program. Between the radio stations you hear the background hiss of naturally occurring radio noise (largely originating in the Earth's atmosphere). You might pick up unwanted interference from nearby faulty electrical equipment. But the desired signals, from your favorite radio stations, are obvious when you find them. So it is with SETI. It is the narrow bandwidth signals that one suspects ETI will use to signal with that are being sought. But just how narrow are these bandwidth signals? In fact, the frequency of extremely narrow bandwidth signals would be spread out slightly during their passage through the interstellar medium by an effect called *dispersion*. Dispersion means that it is pointless concentrating the detection bandwidth in less than about a tenth of a hertz, and this is a pretty good guess at a sensible lower limit for the narrowness of bandwidth best used for SETI.

All SETI surveys are looking for the "carrier" wave on which some signal might be placed. When you tune into your radio, you adjust the dial across a range of carrier frequencies typically of about 0.5 MHz to several tens of megahertz. The music or voice transmission then "modulates" the carrier wave used to carry the signal from the point of broadcast to your receiver. For terrestrial radio, various types of modulation are used. Conventional radio transmissions use either "amplitude modulation" (AM), where the amplitude of the radio carrier signal is made to vary in accordance with the sound signal to be transmitted, or "frequency modulation" (FM), where the frequency of the carrier is varied with the signal to be transmitted. AM radio is simple but susceptible to noise interference; FM is more complicated but noise interference can be filtered out. Thus, FM is used for high-quality stereo radio and also for the transmission of television signals. The third type of modulation is called *pulse-code modulation* (PCM), and is used for the transmission of digital signals (where information is coded as numbers written in binary form). Increasingly information is being transmitted in digital form.

Once a SETI carrier is detected, the story really begins. Any modulating signal will need to be recovered (which is likely to require different equipment), and the message decoded. What a challenge! SETI was never going to be easy.

Most SETI surveys try to limit the channel bandwidth to no more than 1 Hz, while some are moving toward bandwidths as narrow as the 0.1-Hz limit identified earlier. And these narrow bandwidths emphasize the scale of the technological challenge. The SETI radio window covers 10 billion channels of 1-Hz bandwidth! Remember that the Ozma receiver deployed a single channel, and Ozpa deployed 384 channels (an impressive achievement in its day). What was actually needed was a receiver that could simultaneously listen to many millions, if not billions, of channels simultaneously! This is quite beyond the comprehension of our everyday experience, where we listen to radio stations one at a time, or "channel surf" between a few tens of television stations looking for anything other than a repeat episode, or a talk or game show. SETI was demanding the technology to listen simultaneously to millions or billions of channels, and to analyze the outputs of these channels several times a second. Without such technology, there could be no serious prospect of finding ETI. One could have as much time as one wanted on massive telescopes, or as much funding as one desired, but without at the very least a million-channel capability, one would merely be playing at SETI. Drake realized this, and that was the reason why after Ozma, despite his continued passion for SETI, he went off to do other things until the technologists could give him a million-channel capability.

Cometh the hour, cometh the man. Paul Horowitz is an electronics genius. After attending a lecture by Drake in 1969, he became infected with the SETI bug. But he needed funding to develop the million-channel receivers he recognized were needed by SETI. He got it from an improbable source. Carl Sagan was joint founder of an organization called the Planetary Society, which has 100,000 fee-paying members keen to learn about planetary exploration and the search for life elsewhere. The inspired patronage of the Planetary Society (and for a time also of NASA) enabled Paul Horowitz to develop advanced receiver technology for SETI.

Horowitz's first observations involved the searching of 185

nearby stars from the Arecibo observatory (where Drake was then the director). During 1981 and 1982, Horowitz visited NASA's Ames Research Center where he developed a receiver known as "Suitcase SETI." It had a 128,000-channel capability, which represented an incredible advance in the technology for that time, but analysis had to be performed off-line. Suitcase SETI had its only visiting performance at Arecibo in March 1982 when it searched 250 stars. Back at Harvard, it was renamed Sentinel, and (despite its impressive portability) served out its days from 1983 to 1985 undertaking a dedicated SETI survey on the 84-foot radio telescope at Harvard's Oak Ridge Observatory.

But Horowitz had now set his sights on a more impressive goal. If SETI needed the capability of multimillion narrow band channels to do serious surveys, then that would be what SETI got. It is a fascinating fact that *ET* produced the financing for this spectacular development (through a donation from Steven Spielberg to the Planetary Society, on the strength of the phenomenal success of his movie *ET* about a lovable alien). Sentinel became META, the Million-channel Extra-Terrestrial Assay, with a then quite staggering 8-million-channel capability. META was used on the Harvard 84-foot telescope, and in 1991 a Southern Hemisphere META II was set up at the Instituto Argentino de Radioastronomia. META II could look toward the center of the Milky Way, and also at the Magellanic Clouds (our nearest galactic neighbors).

SETI was coming up with new challenges, even with META's capabilities. Horowitz decided that there remained a minimum requirement for SETI. A rapid and automatic reobservation of candidate signals was needed. No more off-line processing—and no more long waits for human reaction. If a likely candidate was detected, then it had to be rechecked quickly, which meant automatically since the funding certainly was not available to have an observer always sitting at the ready at the receiver. An improved discrimination against unwanted interference was required, by the simultaneous observation on three receiving beams. A full narrow band coverage of the complete water hole, from 1.4 to 1.7 gigahertz (GHz), would be undertaken. And what did all this require? Not 8-million channels, but a quarter-billion-channel capability. Thus was born BETA, the Billion-channel Extra-Terrestrial Assay. (Horowitz promised colleagues that it would be "beta than meta.") After 4 years of design and construction, BETA was fitted to the Harvard telescope and switched on in October 1995. Horowitz now has his eyes on a receiver of several-billion channels. In terms of the remarkable achievements so far, there must be every expectation that he will deliver such a technical masterpiece.

Making the Most of Your Chances

Stuart Bowyer of the University of California at Berkeley is an enthusiast. Other "e" words come to mind: energetic, (mildly) eccentric (in the nicest possible sense), exciting, and expressive. Science needs larger-than-life characters like Bowyer, and SETI has been blessed to have Bowyer on the team. Best known for his pioneering work in X-ray and ultraviolet astronomy, Bowyer is usually found where the astronomical action is. And Bowyer decided that SETI would be generating its own special form of action.

Bowyer's interest in SETI was stimulated by the Cyclops report. But Bowyer was not going to wait around to see whether funding for this grandiose project was forthcoming. Bowyer is not the waiting type, once his interest has been stimulated. With his colleague Jill Tarter (then a graduate student), he developed an observing technique of "borrowing" other people's observations. Once radio signals have been received and amplified at a radio telescope, there is no problem feeding the same signal to a SETI "spy receiver" as to the primary researchers. Bowyer did not need dedicated telescope time, and he did not need much funding. He rounded up an assortment of out-of-service equipment, and a borrowed computer, and rode "piggyback" on other people's experiments. He called his program SERENDIP (Search for *E*xtraterrestrial *R*adio *E*missions from *N*earby *D*eveloped *I*ntelligent *P*opulations). SERENDIP did not need to apply for its own telescope time, since it operated parasitically on the observing time of others without degrading their observations. Clever. (The name for the program comes from the story "The Three Princes of Serendip" by Horace Walpole, about making discoveries by accident.)

The SERENDIP equipment was first mounted on the University of California's 85-foot radio telescope, and later moved (as SERENDIP II) to the 300-foot telescope at the NRAO. A later manifestation finally made the big time, and was mounted (as SERENDIP III) on the 1,000-foot Arecibo telescope in 1992. SERENDIP III could analyze 4 million channels at once, and like its predecessors operated at a radio frequency of 430 MHz (rather than the favored 1,420 MHz in the water hole). SERENDIP IV operates in the water hole, and can analyze simultaneously 168 million channels (each 0.6 Hz wide).

With the BETA and SERENDIP IV capabilities, SETI was starting to look serious. So where was NASA?

Out of the Ashes

NASA had big plans for SETI in 1990. Their High-Resolution Microwave Survey (HRMS), with Jill Tarter appointed as project director, was to be a $100 million program extending over many years. It had a dual strategy: targeting 1,000 sun-like stars relatively nearby with a 15-million-channel system operating in the water hole (the Targeted Search System), and also an all-sky survey that would repeatedly scan the whole sky for distant ETI able to transmit at much greater intensities (including the type II and III civilizations). HRMS was to use the Arecibo and Green Bank telescopes and telescopes in France and Australia. The SETI Institute was founded (with Drake as president) as a private nonprofit institution to run the program in an efficient and effective manner. Hopes were high. New technology came online. HRMS was given a grand launch at Arecibo on October 12, 1992, the 500-year anniversary of the journey of discovery of Christopher Columbus (an appropriate symbolism). The sky survey was started simultaneously at the NASA Goldstone Deep

Space Communication Complex. In the first few minutes of observations, the data collected exceeded the sum of all SETI data collected in the previous 30 years. Such was the scale of the NASA program.

Observations commenced. "Lights. Action. Camera." And then, "Cut!" In 1993, federal funding for SETI was inexplicably cut by Congress. It was not only the HRMS that suffered. BETA, SERENDIP, and Big Ear lost federal funding.

Politicians tend to support programs that they believe the voters support, and perhaps the SETI scientists had not marshaled sufficient popular support for their cause (although this seems very surprising when measured against the enormous public interest in aliens). However, SETI remained respectable despite the removal of public finance and the privately financed SETI Institute remains to coordinate the remnants of NASA's major SETI project. But this episode of the withdrawal of federal support for SETI highlights the problems associated with funding fundamental research in a free society.

Just because something is scientifically possible does not make it socially acceptable. Just because something is socially desirable does not make it scientifically achievable. And even if an area of research is scientifically feasible, and socially acceptable, can the costs always be justified? These are the paradoxes of science. Scientists cannot avoid addressing both "facts" *and* "values." Although science is thought by most people to be about "facts" (present perceptions of the real world), the values of the real world will always influence what science is able to do. Indeed, societal needs, identified by the value judgment of ordinary people (articulating their opinions through their elected representatives in the democratic societies), often set the research challenges for scientists: "Please find a cure for cancer," "Please discover nonpolluting sources of energy," "Please rid the world of the nuclear threat." The fact that society now places finding a cure for cancer ahead of the challenge of a manned flight to Mars is understandable, but in the 1960s the challenge of placing a human on the Moon took precedence over all other challenges for U.S. technology. Today's value judgments are different from those applicable at the height of the cold war. While science advances, values change; the "moral environment" in which scientists work waxes and wanes in ùnpredictable ways. Our perception is that a majority of people believe SETI to be a noble and valued endeavor, worthy of the investment of taxpayers' money. They are genuinely excited by the prospect of making contact with ETI. Public opinion surveys show that some 80 percent of the U.S. population believes ETI exists, and a similar percentage supports the concept of SETI. Normally political pragmatism reflects public beliefs, but not, it seems, in the case of SETI where public support proved insufficient to impress cost-cutting politicians.

By supporting all fields of science, a nation is creating the intellectual environment within which all research can flourish. Young people are often attracted into science by the glamour subjects, such as particle physics and astronomy, and they then move on to more applicable fields of endeavor. SETI is a glamour subject for the young, and is proving to be a powerful educational tool to introduce young people to the wonders of the universe. (The educational material produced by the SETI Institute is outstanding, and the Planetary Society also does an excellent job. But even more needs to be done, to counter the antiscientific trivia fed to a gullible public by the tabloid press.)

Following the removal of NASA funds, Drake moved quickly to secure the jobs of the key scientific and engineering staff at the SETI Institute. Miraculously, enlightened private patronage was forthcoming. However, some tough decisions had to be made. The all-sky survey had to be sacrificed, since it needed NASA's deep space network. The Targeted Search System was to be given precedence, looking at 1,000 nearby stars in the frequency range 1 to 3 GHz with a resolution of 1 Hz. The program was rechristened Project Phoenix, after the mythological Egyptian bird that rose again from the ashes—in this case, SETI was rising from the ashes of the loss of federal funding. Four major funders emerged from the high-tech industries: David Packard and William Hewlett, of the Hewlett-Packard company; Gordon Moore, the founder of Intel; and Paul Allen, cofounder of Microsoft. They have all put in a million dollars for starters and have pledged millions more. And Bernard Oliver, vice president of

Hewlett-Packard and a SETI pioneer, left a generous endowment of many millions of dollars to the SETI Institute when he died in 1995. Charitable foundations have also invested in Project Phoenix, as has the general public.

Despite the NASA cuts (and withdrawal of funds from all other federal agencies), with private funding the Targeted Search System was in operation at the Parkes radio telescope in Australia by February 1995, just a month later than originally planned. Then after it was upgraded, which took 6 months, it was mounted on the 140-foot telescope at NRAO for 18 months of observing. In 1998 it returned to Arecibo. Project Phoenix is a remarkable story of scientific commitment in the face of funding failure. The key scientists could have given up on SETI, and moved on to other fields of research still able to attract federal funding. But such was their belief in the importance of the quest, the power of their technology, and the likelihood of eventual success, that they were prepared to press on through seeking private funding. What impressive self-belief!

Because Project Phoenix uses the world's largest radio telescopes, has confirmation from a remote telescope, uses near real-time data processing, and has a rapid confirmation strategy, it is the most comprehensive and sensitive SETI survey ever undertaken. However, it is still only looking at just a thousand stars, in a very small portion of the Milky Way Galaxy (out to a distance of about 200 light-years). And if ETI civilizations in the Milky Way are measured only in thousands, rather than the hundreds of thousands or even millions argued in the past by the SETI optimists, then it is not obvious that Project Phoenix will find anything between now and the year 2001, when the present survey is due to be completed. We hope we are wrong. Each of the 1,000 stars currently being surveyed by Project Phoenix represents a long shot of a million to one for finding ETI. Nevertheless, Drake remains optimistic that ETI will be detected before the turn of the millennium. Many others among SETI's true believers (among whom we most certainly count ourselves) are less optimistic, and are already thinking ahead to a post-Phoenix era. If to find ETI it proves necessary to survey a million stars within 1,000 light-years, then a Phoenix-type survey could take many, many decades rather than just a few more years. Get ready for what might become the long haul.

Bring on the Amateurs

When the U.S. Congress withdrew support from SETI, one of those who took positive action was an amateur radio enthusiast named Richard Factor, the president of a New Jersey electronics company. He formed the SETI League, a group of amateur radio astronomers who would help fill the void left by NASA. What could the amateurs do? Individually, they could do next to nothing in terms of surveying for ETI. But what about a worldwide army of 5,000 amateurs, dividing up the sky and the frequency band? Now their contribution could start looking serious.

Amateur SETI can be carried out for a few thousand dollars, to buy a satellite dish to use as a small radio telescope, a microwave receiver, and a computer to run shared detection software. The SETI League hopes to have assembled their army of thousands of observers by 2001. They still have a long way to go, but their enthusiasm is to be greatly admired. They should be encouraged by the knowledge that amateur astronomers have made very significant contributions to science, with the discovery of comets and supernovae, for example. Once their numbers are sufficient to embrace a sensible survey strategy, then things could really start to happen.

SERENDIP is involving amateur enthusiasts through an intriguing program called SETI@home. SETI@home uses personal computers connected through the Internet. The personal computers are fitted with a special form of "screen saver" software. When the user takes a break from using the personal computer, instead of a conventional screen saver using processing capacity, the personal computer is linked via the Internet to a SERENDIP database. Different types of data analysis will be tried, working through what it is hoped will be a vast network of 100,000 or more personal computers. Users will be able to see on their screens how their computer is impacting SERENDIP data analysis. Perhaps the first faint whisper from ETI will be heard by an enthusiastic amateur taking a break from a computer game!

The Planned Response to Any ETI Detection

SETI scientists have agreed on a protocol to be followed in the case of a genuine contact. The Soviet CTA 102 fiasco must not be repeated. (The SETI League will also observe the protocol.) The significance of any confirmed contact will be so profound for the whole of humanity that any announcement will have to be handled with care.

The protocol has several essential steps: a verification of the detection by the discoverer and collaborators; the establishment of a network of responsible researchers to provide confirmation and continued tracking; and an announcement to the secretary general of the United Nations and to various international scientific organizations.

Only then will the media be informed, with the honor of the first public announcement being given to the discoverer. What an amazing moment that will be for the human race. In informing the public, care will need to be taken to ensure that the nature of the SETI detection is understood. Misled by decades of media speculation about flying saucers and alien visitations, a SETI discovery will need to be explained with great clarity and accuracy so as to avoid public alarm. Until there has been appropriate international consultation, no response to any contact signal should be made.

The movie *Contact* was entertaining theater. The events surrounding the first real ETI contact are likely to be somewhat more business-like, but no less dramatic.

Deception

There are no "rules" that control the way scientists carry out research, although they are guided by the loose framework of the scientific method. There are no formal regulations for the science profession, other than those that relate to such matters as health and safety. The profession is essentially self-regulated, through the belief of scientists in the merit of their task, by their commitment to the established scientific method, by the care with which they formulate ideas and carry out their experiments, and by their willingness to have their results scrutinized by their scientific peers through publishing them in learned journals.

Fortunately, it is rare for scientists to break faith with their follow researchers by deliberately reporting results that they know to be fraudulent. But scientific fraud does occur occasionally. When it does, the world of science is quick to react. The insistence on verification by the SETI research community provides some protection from possible fraud, as does the discipline of the SETI protocol.

Fraud, deception, "cooking the results," and the selective use of data have no place in research. Fortunately, this has been firmly recognized by legitimate SETI researchers. The SETI community has taken great care to ensure that all the safeguards are in place to protect their science from fraud and deception. Frank Drake was once commissioned by a Hollywood movie producer to show how an "ETI signal" could be fabricated (it was to make a great plot for a screenplay, showing how SETI could be duped). After careful thought, Drake concluded that it would be impossible to produce a spoof signal that would fool the serious SETI programs. The promised consultation fee from the movie mogul was never forthcoming.

In October 1998, Drake's confidence that genuine SETI could not be duped was admirably demonstrated when the SETI Institute quickly unmasked an attempted deception. A claimed ETI detection from a star called EQ Pegasi was posted on the Internet. A signal trace was shown, and at first sight the report seemed plausible enough. It was posted anonymously, but then the "discoverer" did reveal an identity (later shown to be bogus) claiming that he was an engineer working for a British communications company, doing some amateur SETI work on one of his company's satellite antennae. The British Broadcasting Corporation (BBC), usually a reliable source for scientific breakthroughs, announced the "discovery," giving it some credibility. The SETI Institute and other SETI professionals were quickly able to demonstrate that the detection was a hoax. The hoaxer had done his homework, and on first sight his report appeared worthy of investigation. But once the professionals checked it out, some pretty obvious errors were discovered. SETI can do without such distractions, but at least the incident demonstrated that Drake was correct in his assertion that even the most sophisticated of SETI hoaxes would soon be found out.

Observation

The SETI systems of today are 100 trillion times more powerful than the simple pioneering surveys of the 1960s. Yet ETI remains just as elusive now as it did then. Why should that be? Could it be that ETI civilizations are not as commonplace in the Galaxy as some would have us believe?

3. Absence of Evidence

"There is something fascinating about science. One gets such wholesale returns of conjecture from such a trifling investment of fact."

Mark Twain

After almost four decades of SETI, absolutely nothing that is certain and verifiable has been found. But absence of evidence for ETI must *not* be interpreted as evidence of absence. Extraordinary claims, such as the claim at the heart of SETI that intelligent life exists elsewhere in the universe, require extraordinary proof. And no such extraordinary proof exists, yet! Some proponents for ETI still rely on tantalizing, but unverified results (such as the Wow signal, and the Martian fossil microorganisms) and on forced logic ("it happened once, here on Earth—so is it really plausible that in a cosmos of unimaginable vastness and age it hasn't happened again?"). Maybe the Wow signal was merely unexplainable interference, and perhaps we should not take too seriously the tantalizing possibility that it and similar events were genuine, albeit transitory, signals from ETI. Scientists often refer to the notion of *Occam's razor*: "Essences are not to be multiplied beyond necessity" (or, in other words, "extraordinary hypotheses are not to be invoked until all ordinary ones have been conclusively eliminated"). SETI has been forced to sit on the sharp edge of Occam's razor, since it has faced such a skeptical audience among funders and evolutionary biologists.

Possible Reasons for the Nondetection of ETI

There are several possible reasons why there have been no unambiguous detections of ETI, despite decades of dedicated search:

1. We are genuinely alone in the Universe. ETI has not been detected because, quite simply, ETI does not exist. There might be plenty of slime out there, but no other intelligent beings.

2. ETI does exist, but is sufficiently rare that detection will require further decades of surveying with instrumentation more advanced than that previously used. SETI had better settle in for the long haul, because ETI is likely to remain extremely elusive.

3. ETI has existed in the past, but there are limits to the growth of any advanced civilization so that any period of potential communication will be short. We just were not around when ETI called.

4. ETI exists, but has no desire to communicate. Why should it advertise its existence to potentially hostile aliens?

5. ETI exists (although rare) and is communicating, but the effect of a phenomenon called *interstellar scintillation* (described later) means that we can only expect to detect transient signals that defy confirmation.

6. ETI's advanced technology is based on alternative communication techniques to conventional radio. Radio might just be too primitive for ETIs far more advanced than we are.

7. ETI is based on a different form of life chemistry, and therefore exists in unexpected places and develops technology of unexpected forms.

Each of these possibilities will now be explored in turn.

All Alone

While it seems entirely consistent with all that is known that simple life-forms can easily emerge on suitable planets, the path to intelligence on Earth depended on more than the roll of the molecular dice. A sequence of entirely random events created an unexpected and extraordinary evolutionary pathway that allowed primates to emerge as a supreme species and humans eventually to rule the Earth. We are the outcome of numerous evolutionary accidents; of that, there is no doubt. Humans are the "winning ticket" in the "universal lottery," and on the "all alone" hypothesis there was no second prize

We can force a conclusion that we may be alone by adopting an approach based on the Drake equation, but inserting plausible terms that are likely to be vanishingly small. Remember that the usual form of the Drake equation is as follows:

$$N = Rf_p n_e f_l f_i f_c L$$

where R is the rate of formation of stars that might have planetary systems and that can survive for long enough for the development of intelligent life; f_p is the fraction of those stars with planets; n_e is the average number of planets per solar system whose environments are suitable for life; f_l is the fraction of habitable planets on which life occurs; f_i is the fraction of life-bearing planets on which intelligence evolved; f_c is the fraction of planets with intelligence where radio communication potential develops; and L is the expected lifetime of communicating civilizations.

The insertion of some extra terms could be defended. For example, an extra R term (R_m) could be inserted to take account of the small fraction of new A, F, and G stars with metal abundances comparable to those of the Sun. (It seems that the Sun might have been formed in an uncharacteristically metal-rich interstellar cloud, bearing in mind its distance from the center of the Milky Way.) A value of 0.1 could be appropriate here.

An extra f term (f_j) could be used to recognize that an Earth-like planet benefits from coexisting with a large Jupiter-like planet, which can act as a gravitational attractor for rogue comets and asteroids and thus protect the life-bearing planet from too high a frequency of catastrophic collisions. Again a value of 0.1 could be appropriate.

An extra n term (n_m) would define Earth-like planets with a large moon, able to stabilize rapid variations in the direction of the planet's spin axis and thus provide a stable climate pattern. From the pattern of planetary moons in our own Solar System, a value of 0.2 would seem to be a reasonable guess.

A further f term (f_x) could be used to recognize that global disasters and catastrophes on a time scale of tens of millions of years open up new evolutionary pathways, and can accelerate the emergence of intelligence, but equally can annihilate life. Here we can probably do no better than insert a further 0.1 factor.

And what about inserting an evolutionary term (f_d) that recognizes that both intellect and dexterity are needed to produce communication and space travel technology (highly evolved dolphins with the intellect of Frank Drake are not going to build radio telescopes to search out ETI)? Very few of the Earth's life-forms display the necessary dexterity, and a value even as high as 0.1 might appear to be optimistic. It could legitimately be argued that this factor is already implicit in f_c, but we separate it here to emphasize a point.

A lifetime term L_f could recognize that a civilization may eventually lose interest in interstellar communication, especially if no replies are received, and may spend just part of its technical era (say 0.1) in a "transmit and listen mode."

This list of plausible additional terms is not intended to be exhaustive; it does not try to include the many social factors that could shape the emergence of a technological civilization. Many additional terms could be added, and sensible values for these new factors estimated.

The revised Drake equation with just these additional terms would now look like the following:

$$N = RR_m f_p f_j n_e n_m f_i f_x f_i f_d f_c LL_f$$

Then, taking the sorts of values proposed for the main Drake factors from Chapter 1, and inserting the new factors proposed here, we get the following:

$$N = 20 \times 0.1 \times 0.5 \times 0.1 \times 0.1 \times 0.2 \times 0.01 \times 0.1 \times 0.1 \times 0.1 \times 0.1$$
$$\times 100,000 \times 0.1 = 0.002$$

(Merely for illustration, a value for L of 100,000 years is used.)

Now the product of probabilities produces a number for N very much smaller than 1. Intelligent life, that is, us, then just looks like a fluke. On this type of analysis we are alone, simply because even we should not be here. The various factors that would allow intelligence to emerge, even if simple life is commonplace, appear to be really so unlikely that we may as well call off SETI here and now. There just is not anyone out there to communicate with! On other worlds there might be slime, fungi, and a myriad of microbes, but no ETI. That is what the pessimists would have us believe. And by adapting a Drake equation approach to their pessimistic expectations, they can predict a pessimistic outcome. An alternative approach of the pessimists is to make just one of the Drake factors infinitesimally small. If anyone wants to be really pessimistic about SETI, then the Drake equation can be manipulated to accommodate that pessimism.

The revised Drake equation above is *not* put forward as anything more than a simple illustration of how pessimism can be extracted from the same tool used by those bent on optimism. We are certainly *not* presenting it as a valid alternative to the standard form of the Drake equation. The analysis presented here is merely trying to make the point that it is easy to "talk down" the likelihood of detecting ETI using the same approach used by many to "talk up" ETI's existence. We have already emphasized that the real value of the basic form of the Drake equation is to help focus on the issues that matter, and that it should not be seen as a computational tool.

We need to recognize that the "all alone" line is one taken by many distinguished evolutionary biologists. They note that the proponents of SETI are invariably physicists and engineers, used to dealing with deterministic processes and encouraged by the laws of physics. It is suggested that physicists have little appreciation of the compounded uncertainties of biological and sociological processes. Scientists and engineers are used to marshaling the arguments to justify the expenditure of vast sums of taxpayers' money (to build radio telescopes, particle accelerators, and space stations), but it has been suggested that biologists and sociologists, accustomed to researching with more modest investments, can take a more realistic view of the nonastronomical factors in the Drake equation. Rather than get drawn into such contentious territory, we will explore instead the intriguing arguments presented by Ernst Mayr (the "grand old man" of evolutionary biology) and Jared Diamond as examples of the style of logical presentation used by those who find the Drake equation approach less than persuasive. Both Mayr and Diamond are passionate and persuasive presenters of the subtleties of evolutionary biology, and it is impossible to be other than impressed by the clarity with which they present counterarguments to SETI. We must stress that we do not share their pessimism, but perhaps that is because we are physicists!

The Mayr argument is a relatively simple one. There are at present, at the very least, some 30 million living species. There is fossil evidence that the average life expectancy of a species is some 100,000 years. Thus, over the almost 4 billion years of evolutionary history, it is possible to estimate that perhaps 50 billion different species have existed on a life-friendly planet, Earth. Yet of these 50 billion species, just one, humans, have developed the intelligence and dexterity to produce communications and space technology. One in 50 billion species on Earth has evolved to produce technology: Should we really expect more than one of the 50 billion Sun-like stars in the Galaxy to have produced a technological civilization? On the Mayr approach, the statistics for ETI do not look compelling.

The Diamond argument is a variation on the Mayr theme. He looked at the simple example of a woodpecker, and asked, How difficult was it for evolution to produce a woodpecker? Not very difficult, it turns out. The evolutionary traits are easy enough to identify. But the niche advantages for the woodpecker are very significant, in terms of ready access to food in any season by digging for bugs in rotting wood (and therefore they have no need to migrate). With such advantages, the question arises as to why the woodpecker niche was not filled on continents isolated from the origins of the present species of woodpecker. If woodpeckers had not evolved, then it seems the woodpecker niche, with all its attractions, would never have been filled. Because an attractive evolutionary path has been filled once on Earth, namely by woodpeckers who have impressive feeding and survival advantages, does not mean that it will be filled again, regardless of how great the diversity of life-forms on the Earth might be. Because an attractive evolutionary path has been filled once in the cosmos, namely by humans who have produced communications technology, does not mean that it will be filled again, regardless of how great the diversity of life-forms in the cosmos might be.

On Earth there are some creatures with abilities that can be compared with those of the last common ancestor of humans and chimpanzees, many living in geographical locations little different from Africa when our ancestors first left the trees and learned to walk upright. Why didn't these other creatures move further up the path to civilization? The fact that intelligent beings now populate the whole planet is not because intelligence emerged independently at all habitable locations. It emerged once, out of Africa, and then migrated. As with woodpeckers, so with humans. It is difficult to come up with robust counterarguments to the Diamond hypothesis.

The simple logic of the evolutionary biologists needs to be taken very seriously. However, the natural caution of the evolutionary biologists has fed the pessimism of the opponents of SETI, who have effectively answered for all practical purposes the "Are we alone?" question to their own satisfaction. Having cut themselves off from any message from ETI by their refusal to support SETI, they are "alone" in their isolation. For them, the "all alone" hypothesis is a self-fulfilling prophecy.

Medium Rare

Many lines of argument lead to the conclusion that if ETI exists, then it is rare indeed. Among these arguments are included the failure of four decades of SETI to produce a positive detection, the complexity of the evolutionary path of humans, and the low likelihood of the emergence of the intelligence and dexterity needed to develop communication technology (regardless of the diversity of life-forms in the cosmos).

An additional consideration is the likelihood that any highly advanced ETI, say 1 billion years more advanced than humans, would have the space travel technology to colonize the Galaxy. There is no firm evidence that aliens have ever visited the Earth, so unless we can prove that interstellar travel is impractical, we might be forced to conclude that highly advanced ETI either does not exist or must be extremely rare. (We will look at the Fermi question—If they are there, why aren't they here?— in great detail in Part C.)

SETI has, necessarily, taken a very limited view of the cosmos. Despite four decades of search, the combination of frequencies and places to look have hardly been touched. Although there have been eighty different SETI campaigns since Project Ozma, they have all been relatively low-budget affairs, usually making use of borrowed telescope time and equipment designed for other purposes. SETI has never enjoyed the lavish support of particle physics or space adventure. (Project Phoenix costs about $5 million a year, which is actually a very modest cost as complex research projects go.)

Soviet SETI concentrated on the proposed strong signals from any advanced type II and III civilizations. But it takes an enormous leap in imagination to conceive of the technologies that could harness the energy needs of such advanced civilizations, and to envisage what their communication habits might be.

U.S. SETI has sampled in-depth across much of the preferred water hole frequency band in only about a ten-thousandth of the Milky Way Galaxy. (It is difficult to estimate with any precision the exact fraction of directions and frequencies surveyed, and the ten-thousandth figure may be an extravagant overestimate.) More of the sky has been sampled on a restricted magic frequency basis, and selected nearby stars have been given particular attention. But in terms of systematic in-depth coverage, such is the incredible scale of the Milky Way that a ten-thousandth coverage of all possible combinations of frequencies and directions is all that has been achieved so far.

Philip Morrison's call to arms—"There is no easy substitute for a real search out there, among the ray directions and the wavebands, down into the noise"—was timely and necessary, and remains so. But so far, all available "ray directions and wavebands" have scarcely been scratched.

The more optimistic applications of the Drake equation have suggested a value for N (the number of radio communicating civilizations in the Milky Way Galaxy) as being between 1 million and 10 million. (The 1961 Green Bank conference had suggested an upper limit to N as high as 1 billion!) The fact that there have been no SETI detections suggests that these optimistic estimates are too high by a considerable factor. If N was really as high as 1 million, then even the limited coverage so far of perhaps no more than a ten-thousandth of the possible locations and frequencies should still have brought forth many tens of detections. But the lack of detection to date would certainly not preclude there being thousands of ETI civilizations out there in our own Galaxy (and similar numbers in the billions of other galaxies). Even with such numbers in the Milky Way, many more decades of dedicated search might be needed to make a detection, even with the sophisticated instrumentation now in place.

Many of the outspoken opponents of SETI have assumed that after four decades of SETI, the sky has been fully surveyed. The job has been attempted, and the result has been failure. In reality, the history of SETI has largely been a series of opportunistic incursions into a complex field of research. The early decades of SETI should not be likened to an Apollo program or a Manhattan Project. There has been no massive budget. There has been no careful central planning and control in the U.S. programs, although there has been some in the Soviet program.

There is no predetermined time scale for a successful outcome (akin to the commitment made by President Kennedy to place a man on the Moon, and return him safely to Earth, within a decade). The opportunistic incursions of the SETI pioneers were not wasted efforts. New techniques have been developed. Various search strategies have been tried, and modified or discounted. New players have entered the field; the early decades were dominated by a few dedicated pioneers, but now SETI is being pushed forward by a new breed of young enthusiasts. They can be encouraged in the knowledge that the negative outcomes of SETI to date certainly do not preclude the possibility that ETI exists, albeit that ETI is probably a somewhat rarer species than the more optimistic early proponents of SETI argued for.

To understand the scale of the SETI search that is needed, consider the following imaginary survey (with apologies in advance for its triviality, but it serves its purpose of emphasizing the scale of SETI). Let us suppose that every person on the Earth was to be surveyed for his or her preference in cola brand. There are approximately 6 billion people on the planet; there are about 400 billion stars in the Milky Way. Thus, a survey of all peoples on Earth would be only about 1.5 percent of the challenge of surveying all stars in the Galaxy. Suppose the cola survey was conducted continuously by a team working around the clock, every day of the year, and let us imagine that it takes 10 seconds to ask each person his or her cola preference, and then to make contact with the next person. At this rate, a full survey would take 1,000 years! Of course, this is nonsense; it would take many times longer than a lifetime, and the sample of people being asked would be constantly changing. The way around this would be to take a small representative group of people, and use this sample to estimate the cola preferences of the total global population. An alternative method would be to employ, say, 1,000 coordinated survey teams and cover the full human population in a year. For SETI, both these approaches are being used, running several surveys in parallel and also taking selected samples of stars. But to emphasize the scale of the SETI problem, even our full imaginary cola survey represented only a small fraction of a full Milky Way survey, and it only asked one question ("What cola brand do you prefer?"), whereas a full radio SETI survey has to ask many millions of "questions" (it has to sample many millions of potential narrow bandwidth frequency channels).

To put our imaginary cola survey into the realm of SETI, we would have to imagine the survey teams inviting comments from billions of people on hundreds of millions of products. The SETI survey specialists have to use every trick that technology can offer when tackling a survey of such mind-numbing scale. SETI is on a grander scale than any other scientific survey ever undertaken. Those who expected quick results (affirmative or negative) have not appreciated the true scale of the endeavor.

Limits to Growth

A big unknown in the Drake equation is L, the expected life span of a radio communicating civilization. Some have argued that there are limits to the growth of any advanced civilization so that any period of potential communication might be very short. Three plausible options are often considered. The first is that a technological civilization overexploits its natural resources, or pollutes its environment to an excessive extent, so that it is forced back into a state of barbarism and a technological "dark age." The second option is that a technological civilization develops weapons of mass destruction, and brings about its own demise, or is killed off through uncontrollable pestilence and disease. Finally, the reoccurrence of natural disasters that produce the evolutionary niche for intelligent species to emerge (believed to be an asteroid collision in the case of the demise of the dinosaurs, opening up the evolutionary path for humans on Earth) could eventually close off that niche.

We will look first at the possibility that *the demands on resources of a technological civilization are not sustainable.* Of course on this topic we are in the realms of mere speculation. But our own species can be taken as an illustration of the accelerating demands on resources, and the damaging impact on the environment, as a civilization becomes obsessed with technology.

Space technology, radio communication technology, and

nuclear weapons technology all emerged during just a half-century—a mere blink of the cosmic eye. What will the next half-century bring? There seems to be no bounds on the human imagination in terms of the development and utilization of technology. But the planet cannot sustain the demand on resources an expanding global population is presently making on it. Between the years A.D. 1 and A.D. 1700, the global population is thought to have doubled. In the subsequent three centuries, it doubled again three times, and is now close to 6 billion. It is expected to reach 10 billion about A.D. 2070. The doubling time of the population of Kenya is just 17 years; in Brazil and India, it is just 35 years. Population numbers per se are not a measure of overpopulation; it is the impact of people on the Earth's environment and resources that really matter. The birth of a baby in the United States imposes 100 times the potential stress on the world's resources and environment as a birth in, say, Kenya. Peoples of the Third World do not pursue lifestyles that demand the disproportionate share of the planet's energy and minerals presently expected by people in the industrialized world. There is an increasing use of scarce resources in the Third World, although this is usually to meet demands from the West.

An interesting question is whether technology, which has put such enormous demands on the planet, might in fact be the ultimate cure for environmental and resource problems, rather than merely their initial cause. Is technological development sustainable, or might a technological civilization eventually self-destruct? *Sustainability* means different things to different people. Certainly there is one ("eco-centric") group within the scientific community who would argue that present high-intensity farming methods and overuse of nonrenewable resources cannot be sustained. This group would argue that there needs to be a return to more traditional farming techniques, based on the use of biological fertilizers and pesticides. Demands for energy must be moderated, with a shift being required away from nonrenewable fossil fuels. Aligning themselves, at least in spirit, with the "green" lobby, this group would argue that consumer societies can no longer continue their "throwaway" lifestyles. Planet Earth simply cannot cope with humans undoing in a matter of decades all that planetary evolution achieved over millions of years. Many present technologies that are overly demanding on energy and nonrenewable resources should be discarded. The "eco-warriors" claim that through the reckless overutilization of the Earth's scarce energy and mineral resources, the uncontrolled pollution of a delicate environment, and the unthinking destruction of natural habitats, humans are on a path to inevitable self-annihilation. As for humans, so for ETI? Of course, we do not know, and it is entirely possible that any long-lived ETI has solved the problems of sustainable development. The rate at which we appear to be destroying our own planet certainly makes one pause for thought.

There is a second ("techno-centric") group of scientists and engineers who will argue, with passion equal to that of the first eco-centric group, that sustainability demands further scientific progress to build on the remarkable technological achievements of the past century. This group recognizes that societal attitudes must change, and that greater use needs to be made of renewable resources and the recycling of nonrenewable resources. But their battle cry is "advancement through science and technology," believing that the eco-centric group could set human progress into reverse.

Which group is correct—those who promote sustainability through a "greener" agenda, or those who argue that putting technology into overdrive will produce solutions? Like so many great debates in science, the truth probably lies somewhere in between. An unthinking pursuit of a "greener" agenda could cause more problems than it solves. Sustainability is a real issue, but it does not yet have a real answer. There is, however, the promise that science and technology can continue to advance, giving greater attention to the use of renewable resources and showing increased concern for the environment. The late Gerard O'Neill was a strong advocate for growth through technology, including the development of space technology to utilize the full energy and mineral resources of the Solar System. He estimated that the Solar System's resources would allow at least a further 5,000 years of sustained technical development. What then? Wars fought over diminishing energy and mineral reserves, leading to a return to barbarism, and the loss of technical capability? A long-lived ETI must have faced and solved these problems, and perhaps one day it will advise humans on how to escape the present rush toward destruction of our home planet.

We will now look at the possibility that *technical civilizations self-annihilate through the development of weapons of mass destruction.* The involvement of science in weapons research has always been a contentious topic. The patriotism of scientists is not the issue. The commitment of the scientists who were part of the Manhattan Project to develop the atomic bomb during the Second World War is well documented. However, after the war many of those same scientists actively campaigned against nuclear weapons. Their sense of duty and patriotism had been honored in producing the weapon that helped to end the war, but they understood the full power of what their scientific research had unleashed and never wanted to see that power used again in anger. Nevertheless, the threat that it might be used ensured 50 years of "peace" in Europe, for which the nuclear scientists deserve some of the credit.

Postwar global politics was dominated by the nuclear weapons race between the United States and the Soviet Union (with other smaller players in the nuclear club). "Mutually Assured Destruction" (MAD), a "suicide pact" between the United States and the Soviet Union, was an inescapable consequence of superpower nuclear armament strategies for more than three decades. The acronym was extended to *nuclear MADness* by those who found themselves the helpless hostages of overkill nuclear policies. Despite various nuclear armament limitation treaties, each superpower still had many times the number of weapons necessary to annihilate the other's urban population, with neither having the ability to implement a quick "knockout punch" and so avoid retaliation. The danger that military analysts saw was an attempt at some form of limited use of nuclear weapons, perhaps a minor standoff between the nuclear powers, taking sides in the Middle East or the Balkans, escalating to a point where one party found it impossible to step back from the brink. But what sort of world would be left after a limited nuclear war?

Research has suggested that even a limited nuclear exchange would bring about a global "nuclear winter" of prolonged darkness, low temperatures, toxic smog, and persistent radioactive fallout. Planet Earth would be left uninhabitable for victor and vanquished, combatant and noncombatant alike. The United States and the Soviet Union thus held the key not only to the survival of their own populations, but also to the survival of the entire human race. With the demise of the Soviet Union leading to an end of the cold war, and its replacement with a "lukewarm peace," there have been significant reductions in the size of the massive nuclear arsenals. But other states are now gaining access to nuclear technology, and a growing concern is that just a few weapons in the hands of misguided politicians could yet precipitate a nuclear Armageddon.

During the cold war, the superpowers were investing far more in weapons research than they were investing in civil research. Many civil benefits have resulted from weapons research, and there is now a growing realization that military technology needs to be considered for "dual use" (namely, the intended military application, with possible civil applications also being considered). Some military research must always be confidential (for example, advanced encryption techniques), but tying up "conventional" technology for military use is a somewhat pointless exercise. At the height of the cold war, the United States had a list of technologies that by law could not be exported. The Soviets found ways around the export bans, and the restricted list proved a greater inconvenience to the U.S. allies than to their enemies. With the end of the cold war, many technologies are more freely accessible, but there is a fear that that could allow the emergence of rogue nuclear states or organizations.

One of the more controversial weapons research programs in recent years was the U.S. Strategic Defense Initiative (SDI), popularly known as "Star Wars." The political vision (to rid the planet of the threat of nuclear war) was noble enough; however, responsible scientists knew that the scientific realities meant that the grand vision of detecting nuclear missiles, and shooting them down before they reached their target, was not achievable. For every great new SDI idea, it was a simple matter to think up a cheaper technical alternative to counter it. Despite knowing that SDI was a misuse of science, there were plenty of scientists who were willing to "sup with the devil" if it guaranteed funding for their own research. As a result, many scientific benefits have arisen from SDI research, but the initial total SDI concept was always flawed.

Many advanced weapon systems are now benefiting from the new technologies developed for civil use, especially information technology. Some have imagined future conflicts being won through the Internet rather than on the battlefield; for example, on the threat of hostilities an enemy's information systems could be infiltrated with "killer viruses" that could silence their communication systems and neutralize their control capability. The Allies proved the worth of such a tactic in the Gulf War. Robotic devices can reduce the risk to troops, and the option arises of fighting future battles by "remote control." Fighting by remote control is only one step away from battles in "virtual reality," the idea of which calls into question the whole concept of war. Perhaps the fact that such notions can be contemplated highlights the futility of traditional forms of conflict, and takes us beyond the bounds of science and into the realms of politics and of ethics. Nuclear self-annihilation certainly cannot yet be taken off the "limits to growth" SETI menu.

It was a growing belief that technology would lead any advanced civilization along the path to self-destruction that forced Shklovskii to drop his support for SETI. He genuinely believed that any ETI civilization more advanced than humans would destroy itself, just as the human race was giving every indication of being likely to do during the cold war.

The final argument related to limits to growth is the realization that the *terrestrial and extraterrestrial chaos* (earthquakes, volcanoes, asteroid and comet collisions, a nearby supernova, and so forth) that helped open an evolutionary pathway for humans could just as easily lead to the destruction of life on Earth. We will look in greater detail at these terrestrial and extraterrestrial threats in Chapter 4. The fossil record of mass extinctions is clear enough. Five mass extinctions have wiped out the majority of the planet's life-forms. ETI would be under similar threat. Indeed, it is possible that the external threat of supernovae may be the main reason why ETI is rare.

Supernovae occur most frequently toward the center of a galaxy, and within the intertwined arms of spiral galaxies. Any ETI that evolved in such regions would not survive the mass extinctions inflicted by the explosion of nearby supernovae. Therefore, there might be a "doughnut-shaped" habitation zone in any spiral galaxy (such as the Milky Way) where in regions between the spiral arms (as presently occupied by our Solar System), the occurrence of supernovae is sufficiently rare that ETI can evolve and flourish. Here could be a natural limiter on the frequency of ETI in any galaxy. If intelligent extraterrestrials lie within the central hole of the habitation doughnut, supernovae would destroy them. Outside the doughnut, the heavy-element abundance of stars would be too low for the formation of planets and life-forms. Only in the portion of the ring of the doughnut not sliced by the spiral arms could ETI find safe haven.

Any one of these scenarios, of a return to barbarism because of overpopulation and overutilization of resources, or annihilation by weapons of mass destruction or natural disasters (both terrestrial and extraterrestrial), look only too frighteningly plausible. They certainly suggest that a value for L of 100,000 years might be wildly optimistic. Thus a big question is, Can a civilization develop and sustain political stability? Historical civilizations, such as those of Mesoamerica, Mesopotamia, Egypt, and China, succumbed to political and social instabilities in relatively short periods of time. Past civilizations on planet Earth do not present a very encouraging picture, even if they lacked the capacity for total destruction of the planet. But would we now be here if Genghis Khan had had control of 10,000 ten-megaton hydrogen bombs (or if, in modern times, had Pol Pot)?

We cannot dismiss the possibility that human civilization, and indeed any ETI civilization, might have a limited existence because of the impact of pestilence and disease. Great plagues decimated ancient civilizations, and even modern medicine, with all its marvelous advances, has had difficulty coping with modern plagues such as the acquired immunodeficiency syndrome (AIDS) and Creutzfeldt-Jakob disease.

The "limits to growth" argument is based on the possibility that technical civilizations can arise with relative frequency but are then short-lived because overexploitation of their planet drives them back to primitive barbarism, or political and social unrest (perhaps resulting from disputes over diminishing resources) leads them to nuclear annihilation, or natural disasters bring their premature end. This somewhat negative viewpoint might limit L to no more than, say, a few hundred years. We are not necessarily saying that the intelligent species would face total annihilation, but rather that they would not survive long as a technological civilization to be able to communicate across the cosmos. They could be driven back to a more primitive state before, perhaps, fighting their way back to a sustainable "low" level of technology over a protracted period. Type II and III civilizations would never appear on this basis, so the lack of success of Soviet SETI is immediately explainable. However, perhaps technology acts as a filter, allowing only those civilizations that have developed wisdom as well as intelligence to progress to a long-term future. It would be encouraging to think so.

Antisocial Behavior

If ETI had reached a highly advanced stage, why would it wish to advertise its existence? Robert Rood of Harvard University noted, "The civilisation that blurts out its existence on interstellar beacons at the first opportunity might be like some early hominid descending from the trees and calling 'Here kitty' to a sabre-toothed tiger." In 1974, at the reinauguration of the Arecibo radio telescope after a major refurbishment, a coded message was sent out into the cosmos toward a dense cluster of stars (some 24,000 light-years away). No magic transmission frequency was used, and there was nothing particularly special about the direction chosen. In reality the transmission was little more than a publicity gimmick for the special event. But it was greeted with hostility in certain quarters. British Astronomer Royal (and Nobel Laureate) Sir Martin Ryle was outraged, claiming that humankind might have revealed itself to extraterrestrials of possible ill intent.

Any communicating ETI is likely to be more highly evolved than humans are, and it is as impossible for us to imagine what might form the collective intent of an advanced ETI as it is for slugs in the garden to have an informed opinion on the hostile intent (or otherwise) of humans. However, developing the notion that there may be a natural tendency for technical civilizations to be short-lived through overexploitation of their planet or self-annihilation, it would be encouraging to assume that any long-lived ETI had solved the problems of social and political instabilities and would thus be a peace-loving civilization. Such an assumption is sometimes used by SETI proponents as justification for their search: If only we could communicate with those who have survived the impact of an advanced technical age, then they could advise us on what steps must be taken to guarantee our survival. This is a nice idea but it unfortunately overlooks some fairly important fundamentals: Detection is one thing; decoding messages is something else again; and conducting a question-and-answer session when the time for a response could be decades, centuries, or millennia is somewhat problematic. The response from SETI optimists is that a question-and-answer session would be unnecessary. It is suggested that a highly advanced ETI would be aware of the potential travails of such lesser mortals as ourselves, so that any message they sent would contain guidance on how to promote sustainable development, avoid war, maximize the utilization of energy, and so forth. It is surmised that ETI would transmit the complete text of an *Encyclopaedia Galactica,* providing the secrets for successful evolution to lesser brethren in the cosmos. Decoding would not be a problem, it is suggested, since ETI would surely base any communication on the universal language of mathematics, and code breakers would eventually decipher the messages being communicated. All this would be an exciting prospect, but appears perhaps a little idealistic.

It could be surmised that if ETI broadcasts its existence, then its intentions are likely to be peaceful; if its intentions were hostile, then surely it would lie in wait for others to broadcast their existence. The evangelist openly preaches his or her beliefs; it is the robber who lies in silent wait.

In the Drake equation, a large value for L (the expected lifetime of a technical civilization) is needed to provide optimistic estimates of the number of detectable ETIs. Even if L is reasonably large, why should we assume that ETI will want to transmit narrow bandwidth signals out into the cosmos throughout its technical lifetime? L might be very large, but after transmitting and listening for decades, centuries, or even millennia without a response, wouldn't ETI perhaps conclude that it might be alone? (Here on Earth the U.S. Congress lost interest in searching for ETI in less than four decades; it is very difficult to communicate with those with closed minds.) If a value of L of just 1,000 years is taken for the average transmitting time for ETI (albeit that they might actually survive for 100,000 years or even very much longer), then the prospect for detection falls significantly, even if ETI is commonplace.

Let us suppose that ETI has chosen not to deliberately transmit signals out into the cosmos to alert other civilizations. Could we nevertheless detect it from "leakage" from its television, radio, or radar signals? A cell phone signal from a distance of 300 million miles can be detected by one of the large Earth-bound radio telescopes. But the challenge at stellar distances (measured in tens of trillions of miles) is a different matter, and can be understood by thinking of radar signals from the Earth. The most powerful signals transmitted out into the cosmos are the planetary radar signals from the giant Arecibo radio telescope. (These radar signals can be bounced off the planets and moons in our Solar System, to map their surfaces.) The Arecibo radar could be detected by a similar radio telescope out to distances of about 3,000 light-years, about 10 percent of the distance to the center of the Milky Way Galaxy. Out to this distance there are some 50 million Sun-like stars. But the Arecibo radar has only been in use for 30 years, so its first nondirected pulses would have traveled a distance of only 30 light-years. To this distance there are just a few tens of Sun-like stars. In addition, the Arecibo radar is used for only about 200 hours each year, so the likelihood of spurious Arecibo signals having been detected by any ETI are low indeed; by inference the chances of us detecting spurious radar from ETI are likely to be small. Military radars operate almost continuously, but they are very much weaker than the Arecibo planetary radar and could only be detected out to distances of about 20 light-years.

Television signals from the Earth have been transmitted continuously for some 50 years, and by now are out to about 50 light-years. But at such distances spurious background signals from television are very weak indeed. An Arecibo-sized radio telescope out in space could detect the Earth's background television transmissions to a distance of only about 30 light-years. (At such a distance it would not be possible to tune to individual television shows, so ETI could not pass judgment on the appalling quality of our television programs; it would just be the combined effect of many television signals that would be detected largely as noise.) An "extraterrestrial Arecibo" would see three peaks in the Earth's background television transmissions, from the United States, Europe, and Japan. Every 24 hours these three peaks would rise and set, and from measuring the Doppler shifts of these rising and setting peak signals, ETI could infer the diameter of the Earth. And from measuring an additional Doppler shift component every 12 months as the Earth orbited the Sun, the Sun-Earth distance could be inferred. ETI would not be able to learn about humans by viewing low-quality television shows, but could learn quite a bit about planet Earth nevertheless. The plausibility of using such a technique in reverse for any spurious background television transmission from a nearby ETI has been tested by receiving the Earth's television background signals reflected from the Moon.

But would an ETI more advanced than we are still be using radio and television transmissions through the atmosphere? On Earth, we are increasingly placing signals on laser beams directed down optical cables. This is a more effective and efficient means of television transmission, and cable television is becoming evermore popular. The alternative of satellite television directs signals back toward the Earth, rather than spurious signals out into space. The Earth may grow quiet in the radio bands over the next few decades, as terrestrial television is taken over by satellite and cable television.

The detection of ETI from spurious background radio transmissions does not look too hopeful, but cannot be totally discounted. Type II and III civilizations are envisaged to have such gargantuan energy appetites that anything might be possible.

Twinkle, Twinkle, ETI Signal

The process by which the stars are seen to "twinkle" is given the scientific name *scintillation*. The light from a star is refracted this way and that by moving irregularities in the upper atmosphere, the net effect being that the brightness of the star as we see it varies rapidly. The light from the majority of stars is of constant brightness, at least over reasonable time scales, and above the Earth's atmosphere they do not twinkle. It is just the effect of a turbulent atmosphere that causes the star to appear to brighten and dim in an erratic manner.

Natural radio emissions from stars are also found to scintillate, but the cause of the scintillation is not turbulence in the Earth's atmosphere, but rather irregularities in the tenuous material lying between the planets and the stars. It was in the process of studying this so-called interplanetary scintillation that radio astronomers discovered pulsars by chance. They were expecting to observe the erratic variation in intensity of some radio sources caused by interplanetary scintillation, but instead detected the regular heartbeats of the pulsars.

ETI's signals could suffer from scintillation. The mean intensity of the signal would flicker in brightness, just the way a star twinkles, albeit that the twinkle time would be much slower— minutes and hours, rather than just fractions of a second. But the variation in intensity could be extremely pronounced. A signal from a very distant source otherwise too faint to be detected on Earth could momentarily brighten spectacularly to be briefly detectable.

The Wow signal, or some of the other transient SETI detections, may have been the brief scintillation enhancement of otherwise very weak and undetectable ETI messages. In present-day searches, rapid verification techniques are employed, and so scintillation may be less of an issue than it was in the past. Only signals from very distant ETI (perhaps of type II or III) would suffer significantly, since it is only over vast distances that sufficient interstellar material would be encountered to cause serious scintillation.

Alternative Communication Technologies

Shortly after the discovery of the laser, it was suggested that ETI might communicate by laser light in the ultraviolet, optical, or infrared spectrum. Indeed a NASA satellite called Copernicus actually searched three nearby stars for ultraviolet laser signals in 1974 (none were found). Soviet astronomers, also in 1974, used their 6-meter-diameter optical telescope to search for laser signals from ETI.

If ETI was trying to communicate with lasers, as some have suggested, wouldn't we expect to see the laser beams sweeping the night sky, much as we do for nighttime advertising or entertainment events? Even if finely focused into a narrow beam, laser light from ETI would be spread out in its vast journey through the cosmos, so we could not expect to see ETI laser beams crisscrossing the heavens. Any spread-out laser light would need to be detected with a giant optical telescope.

Laser beams can carry vastly more information than radio waves (hence, their attraction on Earth for cable communications). Therefore, lasers might be very attractive to ETI. But targeting would be a problem. A laser beam directed at Earth would need to anticipate the movement of the Earth during the travel time of the light pulse from ETI's planet to Earth. Such precision targeting would be an enormous technical challenge, quite beyond the capability of present human technology.

Radio waves still seem a more attractive beacon for ETI, but we need to be alert to the laser option and several small-scale laser surveys are being conducted. These optical surveys may yet surprise us, by making the first positive detection of ETI.

Alternative Life Chemistry

It is the availability of water on Earth that makes life possible, because of the effectiveness of water as a solvent able to facilitate various forms of clever chemistry. Life is a very complex phenomenon, so it is hardly surprising that complex molecules are at its heart. Carbon (C) and silicon (Si) atoms are better able than other atoms to provide a skeleton for the formation of very large complex molecules. When carbon combines with oxygen, it forms carbon monoxide and carbon dioxide, both of which are soluble in water, and subsequent chemistry is possible. But silicon combines with oxygen to form silicon dioxide (sand), which is not soluble in water, so chemistry grinds to a halt. Thus, nature favors carbon as the chemical backbone of the complex molecules of life.

The six most abundant elements in the cosmos are hydrogen and helium (from the big bang), and oxygen, carbon, neon, and nitrogen fused in the stars. Helium and neon are inert gases; that is, they do not take part in chemical reactions. Hydrogen, carbon, oxygen, and nitrogen make up 98 percent of the Earth's biological material (the so-called biomass). The chemical choice of life on Earth thus follows the pattern of the cosmic abundance of the elements. Nevertheless, it might be that other forms of chemistry are able to sustain other forms of life. For example, it has been speculated that a mixture of hydrocarbons could act as the solvent for the complex chemistry of an alien world, allowing life to be sustained over a much wider range of temperatures than possible with water-based life. Perhaps ammonia, with some water, could act as the solvent for an alternative alien world chemistry, allowing strange life to flourish at much lower temperatures (to -50 degrees centigrade). Silicate life (with silicon forming the skeleton of the complex life molecules) might be possible. At temperatures close to 1,000 degrees centigrade the medium would become liquid, forming the basis for a possible evolving chemical order in other worlds hostile to water-based life.

This is all mere speculation, and it is difficult to imagine what forms of alternative technology could be produced by intelligent life-forms living in ultracold or ultrahot environments. But it is unlikely to be radio transmitters.

There is a further possibility here—that ETI biochemistry is similar to ours, but spectacularly faster. This could lead to ETI with fulfilling life spans of, say, days (Earth time) rather than decades. Tens of thousands of generations of ETI advancement could then be compressed into a few tens of Earth years, rather than the millions of years of human evolution. "Time" might be a very different relative measure for ETI.

The Effect of Distance

Imagine that two extraterrestrial civilizations commence transmitting radio signals toward the Earth at exactly the same instance. Let us suppose that one civilization is at a distance of, say, 1,000 light-years, and the second at a distance of, say, 2,000 light-years. The signals from the two civilizations, traveling at the speed of light, would reach the Earth separated by 1,000 years, despite starting their cosmic journey simultaneously. This simple example demonstrates the complicating effect of distance and the finite speed of light (or in the case of SETI, radio waves). Although extraterrestrial civilizations randomly scattered across the Galaxy could perceivably start transmissions toward Earth on a reasonably regular basis, the effect of distance could mean that the reception times are clustered, with many detections occurring at certain epochs and long intervals of nondetection at other times. We might just have been unfortunate enough to have developed the technology able to detect ETI during one of the long quiet periods between the arrival of signals from newly transmitting ETI. Even during this quiet period, hundreds (or even thousands) of different signals could still be winging their way toward us.

If ETI lies at more than a few tens of light-years away, then it will be impossible to engage in dialogue within a human lifetime. If ETI lies at a distance of more than a few tens of thousands of light-years, the possibility is that it will not still exist by the time its signals reach Earth; its message to Earth could be akin to a last will and testament. Indeed, if ETI is sparse, then the round-travel time for radio waves to any nearest ETI neighbor may be longer than the typical survival time of any ETI. A dialogue between ETIs would never be possible. Just to illustrate this point, let us suppose that there were just 100 ETI civilizations in the Milky Way. The average distance between them would be about 10,000 light-years. But if L was less than 10,000 years on average, then any ETI would have died off before its transmissions had been received by any fellow ETI.

There is also a problem in trying to anticipate and understand any ETI survey technique. If we try to search our garden at night by lighting a 6-volt bulb, we will not see much. But if we put that bulb in a torch that focuses its light into a fine "pencil" beam, we can sweep the beam back and forth to survey the garden. If we knew exactly where to look, we could hold the torch beam steady over a period of time. For ETI to attempt to illuminate the whole sky would require inestimable power. If it had a particular reason to target Earth (perhaps the detection of ozone in our atmosphere had alerted an ETI to the possible presence of life), then it could target us with signals for a prolonged period. But let us suppose it was scanning the skies with a radio pencil beam, much as we scan the garden with the torch. Its beam might sweep over the Earth only every few tens or hundreds of years. It could be that some of the unconfirmed SETI signals were the transient sweep of an ETI radio "flashlight."

Observation

We favor the many lines of argument that lead to the conclusion that if ETI exists, then it is rare indeed. These include the failure of four decades of SETI to produce an unambiguous positive detection, the complexity of the evolutionary path of humans, and the low likelihood of the emergence of the intelligence and dexterity needed to develop communication technology (regardless of the diversity of life-forms in the cosmos). We will develop these lines of argument further in Parts B and C, to strengthen the "medium rare" case. The issue is so important, it is worth approaching from various directions through our three main questions.

ETI almost certainly does exist. But rather than there being hundreds of thousands or millions of ETI civilizations in the Milky Way Galaxy, as claimed by the SETI optimists, the number is probably much smaller and will be measured merely in thousands.

The challenge for SETI is staggering. Absence of evidence should certainly *not* be taken as evidence of absence. But Occam's razor remains as sharp as ever. The evidence from SETI will need to be confirmed by many, and be unambiguous in its interpretation. Even then, the skeptics will demand more than one ETI to prove that the outcome is other than unique.

What a wonderful challenge! What a noble endeavor SETI is! A bountiful cosmos, advanced technology, and dedicated researchers: Everything is in place.

PART B

The McCrea Question

•

If life in some elementary form that we know about is available on every planet in the cosmos, what is the chance that creatures like humans will evolve elsewhere?

4. Living Planets

"The created world is but a small parenthesis in eternity."

Thomas Browne

What conditions on a planet are required to bring forward intelligent life-forms? A good place to start investigating this question is with Earth, one certain example of a planet that was able to bring forward a great diversity of life-forms, one of which developed an awareness of natural phenomena and the technology capable of seeking out intelligent beings elsewhere in the cosmos. It is worth the effort to look in some detail at how the Earth evolved, and at the characteristics that impacted the development and evolution of life on Earth. Once we understand how life on Earth evolved, and survived dramatically changing conditions, then we will be in a better position to address some of the issues alluded to in Chapter 3 surrounding the nondetection of ETI so far.

We also need to consider the nature of our parent star, the Sun, since it is the Sun that sustains life on Earth—just as ETI will be sustained by its parent star. Thus, this chapter will consider the Sun and how it and its planets were formed, the Earth and the way it has evolved, and the Earth's environment in space. It will look at the nature of the Earth's atmosphere, the nature of the seasons, the impact of the Moon on the ocean's tides, the weather and climate system, and natural catastrophes such as earthquakes and volcanoes. All of these influenced the establishment and evolution of life on Earth, and thus can be used as a model for the factors that could influence the evolution of life on another planetary system in the depths of the cosmos. Just as humans have been shaped by the Earth's properties and changing character, so ETI will have been shaped by its own planetary environment. The chapter will also consider the extraterrestrial phenomena that have impacted evolution and caused mass extinctions of species here on Earth, and presumably, on the home planet of ETI.

Spaceship Earth

On December 21, 1968, the Apollo 8 mission set off from Earth for a Christmas circumnavigation of the Moon, in preparation for the first lunar landing the following year. The first television pictures of the Earth sent back by the Apollo 8 astronauts presented a sobering image for the inhabitants of "Spaceship Earth." The planet appeared as a fragile blue and white pearl, surrounded by the faint blue haze of its atmosphere, hanging in the blackness of space. It has been suggested that the modern environmental movement had its birth in those first images of the Earth from space, as people realized just how delicate and precious their home planet truly is.

The Earth obtains almost all its energy from the Sun. In the formation of the Solar System, some 4.6 billion years ago, the Sun and all the planets were formed from the collapse of a giant interstellar cloud of gas and dust. It is surmised that the collapse of the proto-solar cloud was precipitated by the expanding shock wave produced by a nearby supernova. The death of one star resulted in the birth of another. Dust to dust, ashes to ashes, on the cosmic scale. A beautiful vision of eternity. As the core of the collapsing proto-solar cloud reached the extreme temperatures needed to initiate nuclear fusion reactions, the newly born Sun started to radiate energy.

The collapsing cloud of gas and dust was rotating, so that around the nascent Sun a swirling thin disc of dust was formed. Substances such as iron and silicon, which can solidify at very high temperatures, formed into dust grains in that part of the swirling disc closest to the Sun, to form the basis of the inner rocky planets. The grains of dust lumped together to form particulates. These in turn accumulated into stones, which in turn accreted to become larger rocks, which over time merged as *planetesimals* (primitive small planetary bodies, a few kilometers across). The planetesimals eventually collided to become the rocky inner planets of the Solar System—Mercury, Venus, Earth, and Mars—plus the asteroid belt lying beyond Mars where the collision rate was not quite sufficient to create a planet. The more volatile substances then condensed farther out in the swirling proto-solar disc, to form the massive largely gaseous planets—Jupiter, Saturn, Uranus, and Neptune—and the small icy outcast Pluto, plus frozen debris lying outside the orbits of the planets, manifested as comets. This process of planetary formation, first proposed by the Russian geophysicist Otto Schmidt, probably took about 100 million years.

Debris from the collapse of the proto-solar cloud that did not accumulate into planets can now be seen as asteroids and comets orbiting between the planets. Way beyond the orbit of the outermost planet, Pluto, there is a ring of millions of comets, known as the *Oort Cloud*, the waste material left over from the formation of the solar system. This frozen wilderness can be perturbed by the effect of gravity of the giant planets or a passing star, and some comets subsequently drift into orbit around the Sun. Comets and asteroids should not merely be thought of as space "junk." They have had a profound impact on life, as we will see later. Just as the formation of life on Earth has been shaped by a hostile space environment, the assumption must be that the evolution of ETI has been similarly influenced. The newly formed Earth would have been bombarded by a protracted shower of such asteroids and comets during its early life, with the result that it gained a great deal of energy. A molten core resulted, sustained by the radioactive decay of the heavy elements in the core. (Radioactive elements, such as uranium, have nuclei that are large and unstable; they disintegrate to throw off energetic "alpha" or "beta" particles, so as to achieve a more stable configuration.) The molten material was the source of the Earth's magnetic field, and its early atmosphere. The early bombardment period by massive asteroids and comets punctuated the early evolution of an atmosphere forming by outgassing from the molten core. The massive blast from an asteroid collision would have stripped away the evolving atmosphere, and created a halo of vaporized rock at thousands of degrees. This was hardly the environment where we might expect primitive life to appear, and survive. But it did. The basis for this current understanding of the early life of our planet will become apparent as we proceed through this chapter.

Without the Sun, the Earth would be no more than an ice-covered ball of rock. The Earth's position, lying comfortably within the Sun's habitable zone, has enabled it to evolve with an atmosphere and oceans, and hence to be able to sustain life in the form we understand (water based, with replicating organic molecules). If the Earth was closer to the Sun, its atmosphere and oceans would boil away, leaving a baked desert like the surface of the planet Mercury. If it was at an extreme distance from the Sun, the dearth of sunlight would leave it cold and uninhabitable, like the planet Neptune. To understand our planet, we must understand its interaction with the Sun.

The Sun

The Sun, like any other star, is an enormous globe of gas at extremely high temperatures. At its center, the temperature must be some 20 million degrees; its visible surface temperature is almost 6,000 degrees centigrade. (Never look directly at the Sun. You could risk eye damage!) The Sun contains 99.9 percent of the total mass of the Solar System. The nine planets orbiting it are visible only by reflected sunlight.

Observation of features on its surface reveals that the Sun does not rotate as a solid body. At its equator, the period of rotation is about 25 Earth days; at middle latitudes, this increases to 27 days; at high latitudes, it approaches 33 days. The surface is a heaving site of violent activity. The Sun's visible surface (called the *photosphere*) is pocked intermittently with dark features called *sunspots*. Sunspots are thought to define knots of intense magnetic field, where the surface temperature is constrained to be lower than its surrounds. Sunspots tend to occur in groups. The frequency of occurrence of sunspots waxes and wanes over an 11-year cycle. Explosive ejections of hot gas, in fine filaments, are often seen. These "explosive prominences" tend to occur around groups of active sunspots. Brilliant outbursts, called *solar flares*, are also a feature of the solar surface during periods of high sunspot activity. They are witnessed as outbursts of unusually high-intensity light. Flares eject clouds of high-energy electrons and protons into space, and these can disturb the Earth's upper atmosphere. At high latitudes, picturesque manifestations of the impact of solar flares are the *aurorae*. At lower latitudes, solar flares can disturb radio and television signals. (While light from the Sun takes 8 minutes to reach Earth, the energetic particles from a solar flare take about a day. Thus, when a solar flare is seen on the Sun, interference to radio reception is likely to follow a day or so later.) Giant solar flares (more intensive than anything observed during recorded history) may have impacted evolving life on Earth in unexpected ways, and ETI will have had to evolve coping with energetic flares from its parent star. Some stars display exceptional flare activity, and presumably their planetary systems could not bring forth life in the form we understand it.

A total eclipse of the Sun occurs when the Moon traverses the line of sight between the Earth and the Sun. During the period of total eclipse, the tenuous outer features of the Sun can be seen. Extending for about 13,000 kilometers above the photosphere is the *chromosphere*, a highly structured layer of almost transparent gas. Far beyond the chromosphere is a faint pearly halo called the *corona*. The gas in the corona is very much hotter than that in the photosphere, at some 1.5 million degrees.

The tenuous corona is not bound to the Sun, and there is a steady flow of gas out into space. This outflow of solar gas is called the *solar wind*. The Earth's environment in space is under the continuous influence of the solar wind, just as ETI's planet will be under the continual bombardment of a stellar wind from its host star. An inhabited planet needs a magnetic shield to protect it from the particle wind.

Although we see the Sun in visible light, and feel its heat as infrared radiation, it also emits ultraviolet radiation and X-rays (from the hot corona). These emissions have an important influence on the Earth's upper atmosphere, and potentially on its life-forms. At times of violent solar activity, it also produces strong outbursts of radio emission that can interfere directly with radio transmissions on Earth.

Thus, in looking at the influence of the Sun on the Earth, we should not think merely in terms of its light, which separates day from night as the Earth rotates under its watchful glare, or in terms of its warmth, which causes plants to grow. We need to think of all of the forms of electromagnetic radiation, the continuous solar wind of particles, and the tongues of energetic particles in solar flares. These factors also need to be considered in thinking about the evolution of life on other planets, since other stars produce all forms of radiation in varying measure and generate stellar winds and stellar flares. ETI is the result of the cosmic evolution and chaos in its own neighborhood, just as the destiny of humans has been shaped by the chaos within our Solar System.

The Magnetosphere

The fact that the Earth has a magnetic field has been known since antiquity. For centuries, sailors used the rotation of a freely suspended compass needle in the Earth's magnetic field as an aid to navigation.

Unlike the magnetic field produced by a bar magnet (the type you might use at home, for example, to pick up pins), the Earth's magnetic field is not symmetrical. It has a form of great complexity and variability. The reason for this complexity is the interaction with the solar wind. The Earth's magnetic field carves out an enormous cavity in the wind, referred to as the *magnetosphere*. The particles in the solar wind cannot cross the Earth's magnetic field, but are swept around it. The side of the magnetosphere facing the Sun is compressed into an elliptical shape. On the far side from the Sun, the Earth's magnetic field is drawn out by the solar wind into a long tail like that trailing from a comet. Lying within the magnetosphere are doughnut-shaped belts of energetic particles (electrons and protons) called the *Van Allen belts* (named after their discoverer in the early years of the space era). The energetic particles in the belts follow spiral paths wrapped around the lines of magnetic field, and bounce backward and forward between hemispheres. The particles in the Van Allen belts can cause damage to artificial satellites passing through them. Special measures need to be taken to avoid damage; for example, the electronic components used in satellites need to be hardened against the effects of radiation damage.

The limit of the magnetosphere is defined by the *magnetopause,* the outer frontier of planet Earth where the effect of its magnetic field and atmosphere are swamped by the effects of the solar wind and extended solar magnetic field. Beyond the magnetopause is outer space.

Although the magnetosphere provides a protective barrier, shielding the Earth's surface from direct interaction with the solar wind (other than those particles able to penetrate the magnetosphere at high latitudes to cause aurorae), the indirect impacts can be significant. The severe disturbances to the Earth's magnetic field, caused by the strong gusts in the solar wind generated by solar flares, are called *magnetic storms.* Such major changes in the Earth's magnetic field may result in sudden enormous pulses of electric current being induced in national power grids, causing power blackouts and damage to electrical equipment. Electrical effects have also been known to occur in oil pipelines during magnetic storms, and precautions have to be taken to prevent explosions.

There is evidence that the direction of the Earth's magnetic field reversed at certain periods in its history. During the time of reversal, the Earth's surface, and life on it, would be exposed to the full intensity of the solar wind. It is likely that the path of evolution was disturbed, in unknown ways, during these intervals of field reversal.

Without the protection of the magnetic field, over a long period the solar wind would sweep away the Earth's atmosphere and life would be destroyed. Had the Earth not had a sufficiently strong magnetic field, it is not obvious that life could have formed, or if it had formed, then its evolution would have taken a very different path through having to cope with direct exposure to the solar wind and a more tenuous atmosphere. A planet that produces ETI is highly likely to have its own protective magnetic field.

The Atmosphere

The Earth's atmosphere is the vital shell of life-sustaining gases encircling the planet. Without some form of atmosphere, a planet cannot sustain life, at least life as we know it. The Earth's atmosphere is presently made up of about 78 percent nitrogen, 21 percent oxygen, and 1 percent water vapor, plus a little carbon dioxide and a pinch of inert gases. Life on Earth has evolved to its present state, nurtured by this makeup of the atmosphere, although the origin of life took place when the Earth had a very different atmosphere. Now humans are forcing spurious gases into the atmosphere, such as sulfur dioxide from power stations and carbon monoxide from motor vehicles, disturbing the delicate chemistry of the atmosphere that sustains life in its Earthly manifestations.

The Earth's early atmosphere was composed largely of carbon dioxide and nitrogen, outgassed from the hot molten core of the early planet. There was plenty of water vapor and probably trace amounts of methane, ammonia, sulfur dioxide, and hydrochloric acid, but *no* oxygen. (This cocktail of gases may have been similar to what we now observe belching from volcanoes.) Apart from the water, the Earth's early atmosphere is believed to be not unlike what we now see on Venus and Mars. The rapid outgassing of the core could have produced water vapor. However, it has also been suggested that a deluge of comets, believed to have bombarded the early Earth, brought the water (plus carbon dioxide and prebiotic molecules) to the planet. It has been calculated that the early epoch of cometary bombardment could have brought all the water needed to fill the oceans several times over.

As the temperature of the early Earth cooled, the water vapor in the atmosphere began to fall as rain, perhaps continuously for 100 million years! The Earth had become an ocean planet. And it was in the oceans that the first simple forms of life were believed to have formed. All the rain had another interesting development on the planet. The rain was a mixture of water and weak sulfuric acid (taking the sulfur dioxide out of the atmosphere). The sulfuric acid would have dissolved calcium in the Earth's primitive crust. The calcium would have reacted with the carbon dioxide in the atmosphere, to form calcium carbonate deposited on the ocean bottom. This removed much of the carbon dioxide from the primitive atmosphere.

The evolution of the atmosphere to what we observe today was closely linked to the Earth's evolving biology. This was through a process called *photosynthesis*. Photosynthesis is the process by which an organism can use light from the Sun, plus water and carbon dioxide, to produce carbohydrates, with oxygen being given off in the process. (Ultimately all life-forms on Earth depend on photosynthesis, since it is the chemical process by which the basic foodstuff, sugar, is produced.) Photosynthesizing microorganisms near the surface of the oceans would have flourished in the carbon dioxide-rich environment, releasing copious amounts of oxygen. Oxygen is chemically highly reactive, and it would have been used up initially in oxidizing minerals in the ocean (for example, iron). Only then would oxygen have been released to the atmosphere, to nurture life-forms that had ventured from the oceans to the land. Over a period of 1 to 2 billion years, the level of oxygen in the atmosphere built up, and that of carbon dioxide and methane diminished. An atmosphere able to bring forth a rich variety of life-forms was now in place—thanks to water, a primordial atmosphere rich in carbon dioxide, simple photosynthesizing microorganisms, and plenty of time.

It is the ability of a planet to sustain water in its liquid form and a gaseous atmosphere that allows us to define the concept of a habitable zone for ETI for any particular star system, or at least a habitable zone for life as we know it.

The density of the Earth's present atmosphere decreases rapidly with altitude above the surface. About 50 percent of the mass of the atmosphere occurs within the first 5 kilometers. The atmosphere is divided into four basic layers—the *troposphere*, the *stratosphere*, the *mesosphere*, and the *thermosphere*—and into three important subdivisions—the *ozone layer*, the *ionosphere*, and the *exosphere*.

The troposphere is the site of the Earth's weather system. It extends from the Earth's surface to an altitude of about 8 kilometers at the poles and 16 kilometers at the equator. It contains three-quarters of the mass of the atmosphere and, very importantly, the components that contribute to weather, such as water vapor, smoke, dust, clouds, and winds. The temperature of the troposphere varies from the average 13 degrees centigrade at the Earth's surface, to about -50 degrees at the top of the troposphere. The air in the troposphere is continuously being mixed by a process known as *convection*, warm air rising in one location and colder air sinking back at some other location.

Above the troposphere lies the stratosphere. Very little water vapor gets to such altitudes, and clouds rarely form. There is little vertical movement of air, although there is significant horizontal movement. (Frequent travelers on long-distance flights will be familiar with the phenomena of "clear-air turbulence" and "jet streams," which characterize the horizontal movement of air at stratospheric altitudes.)

Within the stratosphere lies the all-important ozone layer. Ozone is a form of oxygen molecule containing three rather than the conventional two atoms of oxygen. It is formed when radiation from the Sun causes normal oxygen molecules to dissociate into two oxygen atoms, one of which then combines with a normal oxygen molecule to form ozone. Absorption of ultraviolet radiation from the Sun heats up the stratosphere, so that at an altitude of about 50 kilometers the atmospheric temperature has risen to 0 degrees centigrade, before falling rapidly with increasing altitude. The ozone layer plays an important role in protecting life on Earth from the harmful effects of too much ultraviolet radiation from the Sun. A little ultraviolet radiation is acceptable, giving us our healthy summer suntans and producing vitamin D. (Vitamins are a class of organic compound required by living organisms to maintain good health.) However, too much ultraviolet radiation can cause skin cancer, and too much vitamin D can be toxic—hence, the deep concern that certain chemicals released into the atmosphere through human activity (especially chemicals called *chlorofluorocarbons*, traditionally used as cooling agents in refrigerators and as propellants in aerosols) can be swept up to high altitudes around the poles by the atmospheric circulation system, to destroy ozone and punch holes in the protective ozone layer. Already ozone "holes" with ozone densities below 50 percent of normal levels have been detected at high latitudes. Steps are being taken to limit the use of chemicals that will destroy the ozone layer; the need is now urgent if a tragic rise in the incidence of skin cancers is to be avoided. Life as we know it could not have evolved on the Earth without the protection of the ozone layer to filter out harmful levels of ultraviolet radiation; indeed had early life attempted to venture from the oceans prior to the buildup of oxygen in the atmosphere, it could not have survived the effects of unfiltered solar ultraviolet radiation. The evolution of the Earth's atmosphere, and that of its life-forms, were strongly interconnected, and this will certainly also have been the case for the worlds with ETI.

It is concern about human impact on the Earth's atmosphere, and by analogy ETI's impact on its atmosphere during periods of sustained technical development, that perhaps indicates a limiting effect on the factor L in the Drake equation. This then does call into question the prospect that any ETI can maintain a sustainable development to progress from type I civilization status to type II or III. If only we could make contact with intelligent extraterrestrials, their signals might just reveal how they cope with the challenges of sustained technical development.

Beyond the Earth's stratosphere lies the mesosphere, and then the tenuous layer of hot gas called the thermosphere. The exosphere starts at about 300 kilometers. Within it, the density of molecules is so low that they rarely collide with one another. Some of the more energetic molecules (mainly hydrogen) have sufficient energy to escape the atmosphere entirely, and are lost into space; hence, the name *exo*sphere. The loss is not excessive; less than a kilogram of hydrogen is lost each second, and is in part replenished by protons from the solar wind. Certainly, on the time scales of human existence, the atmospheric loss can be ignored.

Within the thermosphere the atoms can have their electrons removed by intense ultraviolet radiation from the Sun. Atoms that have lost or gained electrons are called *ions*. This region where ions have been created is called the *ionosphere*. The ionosphere acts like a mirror for radio waves up to medium frequencies (high-frequency radio waves can penetrate the ionosphere). For this reason, the ionosphere plays an important role in allowing the transmission of radio waves beyond distances limited by the horizon defined by the Earth's curvature. The ionosphere can be disturbed by solar flares, which thus affect long-range radio transmissions and also radio astronomy (another irritation for SETI).

The presence of an atmosphere around the Earth, in addition to supporting life and driving a climate system, produces several interesting effects, including *scattering* and *absorption*. The light from the Sun is essentially white; that is, it is made up of the various colors of the rainbow—red, orange, yellow, green, blue, indigo, and violet. When standing on the Moon, which has no atmosphere, the Apollo astronauts would have seen the Sun as a white globe in a black sky. However, light is scattered by the molecules and small dust particles in the Earth's atmosphere. Blue light is scattered most, and red light is scattered least. Hence, when we look up into the sky on a clear day, we see the blue wavelengths of the Sun's light scattered in all directions by the atmosphere. The sky is blue because the blue wavelengths of sunlight are scattered by the atmosphere toward any observer. Since the blue light can be scattered many times, the sky becomes bluer the farther away from the position of the Sun one is. And the sea appears blue because the scattered blue light from the atmosphere is mirrored by the sea's surface. The stars are not seen during the day, since the scattered blue light from the atmosphere blankets out their feeble glow. They, of course, are still there. Without an atmosphere, the stars could be viewed day and night and the sky would be black, even when the Sun is high in the sky.

Scattering also explains why the rising and setting Sun appears red. When the Sun is low on the horizon, its white light has to pass through a greater depth of atmosphere, in which much of its blue light is scattered, leaving the image of the Sun diminished in brightness and looking red. It is not that the Sun's image contains more red light; it contains less blue light and therefore appears red.

When radiation from the Sun passes through the atmosphere, some of it is absorbed. For example, molecular oxygen absorbs red light, and carbon dioxide and water vapor absorb infrared radiation, reducing the heat from the Sun reaching the ground. Heat emitted by the Earth cannot escape into space, since the carbon dioxide and water vapor in the atmosphere are opaque to infrared radiation. The atmosphere thus provides a thermal blanket for the planet, the so-called *greenhouse effect*. (In a domestic greenhouse, sunlight can pass easily through the glass to power the plants' photosynthesis, but the heat from the plants finds it difficult to escape through the glass. By analogy, the similar phenomenon in the atmosphere gained the name "the greenhouse effect.") The planet Earth and its life-forms have evolved over the eons, with a delicate balance being achieved between radiation received from the Sun and that trapped within the atmosphere. Now there are serious concerns that the atmospheric greenhouse effect may get out of control. More heat radiation is being generated by the industrialized world, turning fossil fuels into other forms of energy that feed heat to the atmosphere. In addition, the atmosphere is being polluted with certain gases (called *greenhouse gases*) that trap this additional heat radiation within the atmosphere. The effect is that the average global temperature is gradually increasing. There is widespread concern that this "global warming" will partially melt the polar ice caps, raising the level of the oceans, and flooding low-lying land areas. The time scale for a significant impact from global warming remains somewhat uncertain, but the situation is sufficiently serious that conferences of the world's leaders have been called to put in place firm measures to control the emission of greenhouse gases. Some industrialized countries are having difficulty facing up to the measures necessary to reduce the emission of greenhouse gases. Unsurprisingly, Third World countries seeking to advance through industrialization are reluctant to take on the extra financial burden of controlling atmospheric pollution if wealthy Western countries preach moderation but are not prepared to take tough action themselves. It is going to be a close call for planet Earth, despite its robustness to abuse, if the present generation of human inhabitants is not prepared to take their temporary stewardship more seriously. Life on Earth is under threat. For humans, the L in the Drake equation might be exceedingly short.

For life on Earth, the greenhouse effect has been at an acceptable level at times when it mattered. When the energy from the Sun was only some 75 percent of its current value during the early period of its life, the Earth's atmosphere sustained a more intense greenhouse effect and therefore kept the planet's surface warm despite the lower solar radiation. As the Sun brightened, the Earth's atmospheric makeup was changing with a reducing greenhouse effect. Thus, the increasing intensity of the solar radiation did not result in any significant increase in global temperature.

For the planet Venus, the greenhouse effect has been so extreme as to have raised the temperature of the surface to levels unable to sustain life as we know it. The emergence of ETI on any distant planetary system will have depended on the extent to which the tendency toward an extreme greenhouse effect emerged. If there is any uncontrolled greenhouse effect, such as that suffered by planet Venus, on a planet with ETI, then it is farewell to ETI. Perhaps even more sobering, if there is any uncontrolled greenhouse effect on Earth, then it is farewell us. We are not joking!

Weather and Climate

The Earth's weather and climate system is governed by the subtle global energy balance. Of the incident radiation from the Sun, about 25 percent is reflected from the atmosphere, a similar amount is absorbed by the atmosphere before reaching the Earth's surface, about 45 percent is absorbed by the Earth's surface (land and oceans), and about 5 percent is reflected from the surface. Some of the energy absorbed by the ocean is lost again in evaporation. Of the infrared radiation reemitted by the Earth's surface, some 88 percent is trapped within the atmosphere by the greenhouse effect.

The atmosphere is in a state of continuous change, and this affects the varying patterns of weather. The weather is driven by two basic principles: hot air rises, and air flows from regions of high pressure to regions of low pressure. The weather can respond to these two principles at the microclimate level (for example, hot air rising above a bitumen road causing the illusion of shimmering images), right up to the global scale (for example, giant cyclones sweeping continents). On the global scale the restless atmosphere results in a drift of hot air from the tropics toward the poles, with cold air flowing from the poles toward the equator. Without this energy redistribution by the atmosphere, the equatorial regions would become increasingly hot, and middle latitudes increasingly cold. Complicating factors in the atmospheric circulation and weather patterns are the rotation of the Earth, the fact that the sea acts as an enormous sink of heat that is given up more slowly than the heat absorbed by the land, the ocean currents driven by the winds, and the existence of the giant land masses and mountain ranges.

Whereas weather changes at any location day by day, *climate* is the long-term variations resulting from the cumulative effect of the Earth's weather patterns in different regions of the globe. Climate is reflected in seasonal variations and trends, such as average temperature and rainfall. The global climate system is very complex, and despite significant advances in recent years, scientists' present knowledge of the system is still only rudimentary. The reasons for climatic variability and change are not well known, although it is apparent that many influences are at work. Potential sources of variability clearly exist within the climate system itself, the principal components of which are the atmosphere, the oceans, the land, the *biosphere* (the Earth's system of life-forms), and the *cryosphere* (the Earth's ice and snow cover). In addition, several processes outside the climate system might force changes. Possible influences of a terrestrial origin are volcanoes, various human influences on the environment that produce greenhouse gases (such as deforestation and the excessive use of fossil fuels), and the ponderous movement of the continents, which over the time scale of hundreds of millions of years affect climate through their influence on the depth and shape of the oceans, the extent of ice cover, the height of mountains, and so forth. In addition to these terrestrial influences on climate, extraterrestrial influences such as long-term variations of the Earth's orbit around the Sun also have an effect (and seem to have accounted for the periodicity in the occurrence of the Ice Ages). Whereas we have no control over the daily vagaries of the weather, or geological influences on climate, human activity can, and is, affecting long-term climate trends. The evolution of life on Earth, and its fight for survival, has had to cope with spectacular variability of the climate. It is reasonable to assume that ETI will also have had to cope with the vagaries of climate on its own world, since the evolution of the factors that drive climate and those that affect the evolution of life are so closely linked.

The Earth's Motion in Space

The notion that the Earth rotates about an axis once every 24 hours can be traced to the ancients. Cicero wrote that the philosophers of his time believed that all objects in the heavens were fixed in position, and their apparent movement resulted from the rotation of the Earth. This view did not go unchallenged. Ptolemy argued that the Earth was fixed at the center of the universe, and all other bodies rotated around it.

It was not until the Renaissance, with the work of Copernicus, that the Ptolemaic view was finally put to rest. The view that Earth spun on an axis, and was in orbit around the Sun, finally gained wide acceptance (albeit still being challenged by the conservative theology of the age).

A familiar manifestation of a rotating Earth is the deflection of air and water currents. On a stationary Earth, air and water currents would converge radially toward a low-pressure area. On a rotating Earth, air and water currents circulate counterclockwise in the Northern Hemisphere and clockwise in the Southern Hemisphere, such circulation being artistically demonstrated in the cloud patterns photographed from satellites, and pictured each evening during television weather forecasts.

The Earth rotates eastward around the Sun in an elliptical orbit, once every year. It is sometimes wrongly believed that the seasons are related to the Earth's distance from the Sun. Actually the Northern Hemisphere of the Earth is nearest the Sun in winter! The seasons result from the fact that the Earth's equator is inclined at an angle (approximately 23.5 degrees) to the plane in which the Earth orbits the Sun. As the Earth rotates around the Sun, the North Pole is always tipped at 23.5 degrees to the plane of the orbit. When the North Pole is tipped as far toward the Sun as is possible (and bathed in perpetual sunlight), this is the "summer solstice" in the Northern Hemisphere. At this time, the Sun will rise as high above the horizon as it ever will in the Northern Hemisphere during the year. Six months later, when the North Pole is tipped as far away from the Sun as is possible (and in perpetual darkness), this is the "winter solstice" in the Northern Hemisphere. Between these two extremes lie the autumnal and vernal (spring) "equinoxes," when the maximum elevation of the Sun passes over the equator.

Thus, the seasons are shaped by the tilt of the Earth's axis and its orbit around the Sun; if the Earth's equator had been perfectly aligned with the plane of the Earth's orbit, there would not be any seasons. In addition, seasons provide the basis of the annual cycle of behavior of most plants and many animals. On a planet without seasons, namely, a planet with its spin axis exactly perpendicular to the plane of the orbit, life would be very different from the flora and fauna of our understanding. The evolution of ETIs will surely have been influenced as strongly by their seasons as life on Earth is influenced by the beauty and variety of our seasons.

The Interior of the Earth

The Earth's interior is divided into three major parts: the *crust*, the *mantle*, and the *core*. The crust is the familiar outer layer, with a thickness that varies from just a few kilometers beneath the ocean floor, to some 50 kilometers or more below the mountainous parts of the continents. The approximate proportions of the elements in the crust (by weight) are as follows: oxygen, 46 percent; silicon, 27 percent; aluminum, 8 percent; iron, 5 percent; calcium, 3 percent; and potassium and sodium, each about 2 percent. The mantle extends to a depth of 3,000 kilometers. It consists mainly of silicates rich in iron and magnesium. The core is made up of two parts: an outer molten core and a highly compressed inner solid core. The core is believed to be mainly nickel-iron. Quite simply, as the molten Earth formed, the heaviest elements (such as iron) sank to the inner regions and lighter elements (such as calcium) stayed closer to the surface. The temperature of the core is estimated to be about 3,700 degrees centigrade. The initial source of this heat was likely to have been the bombardment of the newly born Earth by planetesimals. In addition, however, there is an ongoing source of heat, the radioactive decay of elements such as uranium, thorium, and radioactive potassium. The fluid in the molten outer core is circulating, due to the Earth's rotation and heat-induced circulation currents. It is this circulation in the outer core that produces the Earth's magnetic field. And as already noted, the magnetic field protects the Earth's surface from the solar wind.

We tend to think of the Earth under our feet as solid and immovable. Earthquakes remind us that this is not the case. The occasional earthquake indicating a local rearrangement of material in the crust hides a steady drift of the crust over the millennia, which dwarfs the effects of even the most dramatic of earthquakes. This ponderous motion is called *continental drift*, or *plate tectonics*.

The startling idea that the continents are "floating" on the mantle, like logs on a lake, was first proposed by the German scientist Alfred Wegener in 1912. He pointed out that the coastlines of South America and Africa matched like pieces of a jigsaw puzzle that had been separated. He proposed that Africa and South America were originally parts of the same giant supercontinent (Pangaea), which had drifted apart. This idea was first met with some derision, but in recent decades fossil data and geological data have supported this general hypothesis. There seem to be about a dozen active "plates" drifting across the mantle. The drift is very slow, only about an inch or so a year. At this rate, the 3,000-mile separation of South America and Africa would have taken about 100 million years.

Whenever a plate pushes against its neighbor, the crust shows signs of geophysical activity, forcing up mountain ranges. For example, the Andes range represents where the South American plate has collided with the South Pacific plate.

What drives the drifting of the continents? The driving force is likely to be heat flow from the hot core. Matter heated at the bottom of the mantle rises up under the crust and spreads out just under the crust; the crust is dragged along by this motion. The sudden release of stresses built up in the crust produces earthquakes. Over the longer term, the continents drift.

Volcanoes are closely associated with plate tectonics. They can occur where two plates are sliding apart, such as the underwater volcanoes along the mid-Atlantic ridge, or they can occur where two plates collide, such as the volcanoes of Japan. They represent breaks in the crust-mantle interface (the so-called Moho), through which hot mantle material can flow. Intraplate volcanoes, such as those in the Hawaiian chain, can also occur. Underwater volcanoes can be of such a scale as to produce a structure that rises above the surface: a new island.

On the surface of the newly born Earth, volcanoes would have belched forth gases and lava, producing the early atmosphere and shaping the early distribution of hot land masses and oceans. This activity was supplemented by the continuing impact of planetesimals.

The destructive power of volcanoes can be awesome. In modern times the improved understanding of volcanoes and good monitoring procedures can predict eruptions, and has thus reduced the number of human tragedies. Nevertheless, a relatively small, unpredicted eruption of the Nevado del Ruiz volcano in Columbia in 1985 killed 25,000 when it precipitated giant mud slides that wiped out the town of Armero. Volcanoes remain one of nature's untamed monsters.

The gas and dust belching from volcanoes can be swept by atmospheric circulation to high altitudes and over vast distances.

These clouds can cut off light from the Sun and affect weather patterns. There is evidence that the massive Krakatoa eruption of 1883 had a severe and lasting effect on the climate. The erupted material might sometimes contain poisonous or suffocating gases. Sulfur dioxide from eruptions can mix with water vapor in the atmosphere, and fall as "acid rain," which kills off vegetation (a phenomenon also associated with emissions from power stations burning fossil fuels). Life on the active planet Earth has had to be highly adaptable to periods of sustained volcanic activity. ETIs will have had to be highly adaptable to sustained volcanic activity on their planets.

The destructive effects of volcanoes need to be balanced by the fact that they produce fertile soil, valuable mineral deposits, and geothermal energy. Volcanic ash weathers with time to produce a rich loamy soil. Underground reservoirs of steam and hot water are often a feature of volcanic regions, and drillholes into these reservoirs allow the rapid release of steam to drive electricity generators. Hot water from such reservoirs can be used for central heating purposes. Some such underground geothermal systems will distill elements such as gold and silver from their host rock, producing rich seams of such precious metals.

Volcanoes and earthquakes have had long-term as well as short-term impacts on life on Earth, and their equivalents on distant worlds will have profoundly influenced the evolution of ETI—of that we can be certain.

There is a series of interconnected effects here, for humans and for ETI. A molten interior of a planet produces outgassing to create an atmosphere, which is good for life. It also drives volcanoes, which can be threatening to life. A molten core produces a magnetic field, which protects life from the solar (or stellar) wind. It also drives continental drift and produces earthquakes, which can be threatening to life. Ozone at sea level is threatening to life. Ozone in the stratosphere protects life from harmful solar (or stellar) ultraviolet radiation. Radioactive material in the Earth's core provides energy to drive a dynamic evolving planet. But direct exposure to radioactive material is threatening to life. And so it goes on. For various planetary properties, as far as life is concerned, one can say, "First the good news, ... and now the bad news." The planet and its life-forms are richly interdependent.

Although we do not wish to labor the point about the dramatically changing geological structure of the Earth, it is obvious that continental drift, a changing magnetic field, volcanoes, and earthquakes have impacted the evolution of life and the very survival of certain species. These geological phenomena are essentially sporadic. Thus, evolution cannot find a way around volcanoes, for example, and then move on. A sustained period of volcanic activity may have shaped a particular evolutionary epoch; however, survival of life will always be threatened by such natural phenomena. ETI will be as vulnerable as humans. A living planet is also a threatening planet. The natural phenomena that destroy species, however, can create niches into which other species can move and improve. The natural dynamism needed for biological evolutionary advancement, which we will discuss in the next chapter, is driven by a planet's geological evolution. As it has been and will be for humans, so it has been and will be for ETI, because physical and biological principles are universal.

The Moon

After the Sun, the Moon is the most familiar object in the heavens. It has a special place in the human imagination (think of how many love songs make reference to the Moon!). It is the only celestial body visited by humans so far. It orbits the Earth once every 28 days, at a distance of almost a quarter of a million miles. A moon has more influence on the evolution of life than one might at first expect, and it might just be that a moon or moons are as important for the evolution of ETI as are an atmosphere and magnetic field. The whole "package" of water, atmosphere, geological dynamism, magnetic field, moon, and so forth may be needed for a viable home for any ETI.

The origin of the Moon remains somewhat uncertain. Was the newly forming Earth stretched into a dumbbell shape by the Sun's gravity, with the Moon material fragmenting off as the Earth's satellite? Was the Moon formed as an entirely independent body in the solar system, and then later captured by the Earth's gravitational field? The most plausible explanation is now thought to be that the primeval Earth collided with a truly massive interplanetary interloper, almost a tenth its own size. The incoming object was obliterated in this massive blast, and the debris from the impact formed a disc around the Earth from which the Moon then condensed. This hypothesis has a couple of attractive side effects. The impact would have caused the Earth to spin more rapidly, giving it a relatively short day, and it could have tilted its spin axis, giving the seasons. The exact origin of the Moon remains uncertain, although with moons being a common feature of the planets in the Solar System, the formation process must be reasonably routine. What is certain, however, is that the Moon has interesting effects on the Earth's behavior.

The period of rotation of the Moon about its axis equals its orbital period around the Earth. Hence, the Moon always presents the same visible surface to the Earth. Until spacecraft were put in orbit around the Moon, its reverse side remained a mystery. Because the Moon has no atmosphere, its surface is heavily pocked with craters formed by colliding meteorites. This is especially so on the reverse side where it is fully exposed to impacts from space debris (while on the near side, the Earth provides a partial protective barrier).

The illuminated face of the Moon changes throughout its 28-day orbit. This is referred to as the *phases of the Moon*. When the Moon lies between the Earth and the Sun, its face appears almost dark; this is the *new Moon*. The new Moon rises and sets with the Sun. Over the next 7 days, the illuminated portion becomes a crescent of increasing size, and it rises higher in the sky after sunset each day. It reaches its *first quarter* when one half of its disc is illuminated. Then 7 days later it will lie on the opposite side of the Earth from the Sun; it is fully illuminated as the *full Moon*. The full Moon rises at sunset and sets at sunrise. Then 7 days later, the Moon is in its *third quarter*, and 7 days later the cycle is completed.

Eclipses occur when the Earth, the Sun, and the new or full Moon lie in a straight line. Since the Moon's orbital plane is inclined to the Earth's orbital plane, the new or full Moon usually lies above or below the Earth's orbital plane and an eclipse will not take place. Only rarely does a true alignment occur. These occurrences can be predicted, although in ancient times the failure to predict eclipses meant they were interpreted as portents of disaster. A total solar eclipse lasts just 3 minutes: The Earth's temperature drops several degrees, bright stars can be seen, and the ethereal solar atmosphere can be observed. It is an event of awesome, yet frightening, beauty.

The most dramatic influence of the Moon on the Earth is the ocean tides. For people living near the sea, the rise and fall of the tide is an everyday occurrence; for many shoreline creatures, the waxing and waning of the tide ensures their survival. Tides are caused by the gravitational attraction of the Moon (and to a lesser extent the Sun) on the waters of the oceans. The Moon produces two water bulges on the Earth's surface: one on the Moon side and in line with the Moon, and the other on the opposite side from the Moon. As the Earth rotates, it encounters these regions of "bulge" twice, so that there are two high tides and two low tides each day. The tidal influence of the Sun is much less than that of the Moon because of its very much farther distance from Earth. However, when the Moon, Earth, and Sun are aligned (that is, at full and new Moon), the combined effect of the Moon and Sun produces tides of greater size; these are called *spring tides*, and occur twice each month. Between these extremes are much smaller tides, when the Moon is in its first or third quarter; these smaller tides are called the *neap tides*. Actual tidal patterns are influenced by the shape and slope of coastlines, and the shape and slope of the ocean floor off the coast.

Life started in the sea. Life at the seashore is shaped by the tides. The tides aided the transition from sea life to land life. Quite simple, we are here courtesy of the tides, and therefore, courtesy of the Moon.

It is possible that the Moon played an additional impressive role in the evolution of life not previously supposed. The tilt of the axis of rotation of the Earth with respect to its orbital plane around the Sun is called the *obliquity* of the Earth's orbit. As already noted, it is this tilt that brings the seasons that sustain the annual pattern of life. The tilt changes direction very slowly over very long periods, so that life can adapt to major seasonal variations (such as the onset of the Ice Age). The gravitational interaction between the Earth and the Moon has damped out any wild fluctuations in the Earth's obliquity. It is possible that advanced life-forms can only evolve on a planet that has a large Moon to control massive variations in the planet's obliquity and concomitant wild variations in seasons and climate that would be threatening to advanced life-forms. If this notion is correct, then it could severely limit the incidence of ETI, since while planets within habitable zones might be common, the presence of a dominant moon for planets within a habitable zone might be relatively rare. No moon, no ETI. For sure, this is a highly speculative idea, but it should not be dismissed lightly.

Oceans

Some 71 percent of the surface area of the planet is water, concentrated in five great oceans: the Atlantic, the Pacific, the Indian, the Arctic, and the Antarctic (southern) oceans. In addition to the mighty oceans, there are the "seas" surrounded by land (such as the Mediterranean, Black, and Red seas). The North Sea and the Baltic Sea are merely flooded low-lying areas of continent. The science of the oceans and seas is called *oceanography*.

The floor of the ocean has been mapped extensively by sonar. Sound waves are reflected from the ocean floor, and the time delay from transmission to reception indicates the depth of water traversed. The ocean floor is made up of plains, mountains, and deep rifts; just as with the continents, these are shaped by plate movements. Underwater volcanoes produce the submerged mountain ranges. Rifts are formed where plates collide, and one is driven beneath the other. Near-boiling water belches from deep ocean vents, and the detection of simple organisms around these vents has stimulated scientists' thinking about the origin of life. Darwin proposed that life originated in hot shallow pools on the surface of a primeval Earth. It might be that the origin of life was actually a deep-sea phenomenon.

Sea waves are set up by the prevailing winds wiping across the sea surface. The longer the wind blows, and the stronger it is, the larger the waves are. The effect of the wind is one of the main driving forces of the orderly current systems that link the world's oceans. The other reason for currents is the horizontal and vertical differences in the density of seawater, resulting from differences in seasonal heating of the oceans, the degree of evaporation, and the amount of rainfall. These effects produce currents at all depths of the ocean, the nature of which are also affected by the Earth's rotation. One of the best-known current systems is the Gulf Stream, which brings warm Caribbean waters to the west coast of Britain, and as a consequence a more moderate climate than other land masses at a comparable latitude. Some currents can have sudden and significant consequences. One such is El Niño (the child), caused when warm water from the equator encounters colder water from the Peru current off the coast of Ecuador. Sea life is killed off, and birds dependent on this source of food die. There can be severe effects on the climate also.

Life came from the oceans, and life continues to be shaped by the oceans. We suspect that ETI almost certainly lives on ocean planets.

Gaia

The Gaia hypothesis comes from the ancient Greek word for "Mother Earth." This intriguing and controversial hypothesis was formulated by a British physicist, James Lovelock, and an American microbiologist, Lynn Margulis. The Gaia hypothesis proposes that life on the planet Earth has long controlled the temperature and composition of the atmosphere for its own benefit. Although developed specifically for planet Earth, Gaia provides an intriguing model for any planet producing ETI.

The study of the Earth, its environment in space, and plants and animals and their interaction with each other and with their environment, is usually conducted "bottom up," that is, as a series of unrelated studies of individual species, or of apparently independent physical phenomena. The Gaia hypothesis is a "top-down" approach. It looks at the planet Earth as a whole system, just as one would study a living organism as a whole system. Co-evolution of the organic and inorganic parts of the whole Earth system is at the heart of Gaia. In Lovelock's words:

> The Gaia hypothesis states that the lower atmosphere of the Earth is an integral, regulated and necessary part of life itself. For hundreds of millions of years life has controlled the temperature, the chemical composition, the oxidising ability and the acidity of the Earth's atmosphere.

The Gaia hypothesis simply states that mechanisms of physical and chemical control exist in the totality of life on Earth, and act for the benefit of life. The whole Earth adapts to changes in such a way that life benefits. Lovelock likes to use the analogy of human physiology. For example, the brain can act as a thermostat so that the healthy human body is kept at near constant temperature, despite the wide range of external temperatures that humans encounter. Gaia is presented as "geophysiology" where the "body Earth" has a similar ability to control its condition within reasonably close tolerances beneficial to life.

Consider the level of oxygen in the atmosphere. As already noted, the present high level of oxygen in the atmosphere is thought to have a biological origin. Without photosynthesis in plant life, oxygen would largely be trapped either in water vapor or by minerals that take up oxygen, and the atmospheric level would be considerably lower than the present 21 percent. On the other hand, if atmospheric oxygen had risen above approximately 30 percent at any time, there would have been disastrous fires initiated by lightning. There is also probably a biological sink for oxygen—methane produced by bacteria combines with oxygen to form carbon dioxide, absorbed by plants. Thus, biology appears to sustain the level of oxygen in the atmosphere needed for life, and also controls the level so that it does not rise excessively. Thus, feedback keeps the oxygen level within very tight boundaries.

The Gaia hypothesis has attracted a great deal of controversy since the idea first came to prominence in the early 1980s. It was embraced with enthusiasm by the environmental lobby, which saw it as endorsing the "whole Earth" philosophy they expounded. Some people gave Gaia a pseudoreligious interpretation, which was never intended by its original advocates and served only to deflect serious scientific debate. Where science has looked seriously at Gaia, some of the basic principles have been found wanting, but the Gaia approach has not been entirely dismissed by serious science. Science thrives on controversy, and Gaia ideas will inevitably evolve and be strengthened. As Lovelock himself commented on Gaia:

> In the early days when it was a bit poetic, one thought of life as optimising conditions on Earth for its survival. Now that I understand the theory behind Gaia very much more than I did then, I recognise that this is not so, that it's nothing as highly contrived or complicated as that. There's no foresight or planning involved on the part of life in regulating the planet. It's just a kind of automatic process.

The religious mystique has been removed from Gaia; the potential to use a scientific "top-down" approach to the co-evolution of the planet and life remains. The Gaia hypothesis has increased our awareness of the intricate interplay between the Earth's geological, atmospheric, climatic, and biological evolution. What a fascinating web evolution has woven—for planet Earth and humans, and for planet "X" and ETI.

Extraterrestrial Impacts

The history of planet Earth is a turbulent one. Many dramatic changes have occurred solely or largely through "internal" processes, such as continental movement and mountain building through the process of plate tectonics. However, there is a growing awareness that external factors may be just as important in explaining the many abrupt changes on the Earth over its evolutionary history, disruptions both of climate and of the Earth's biological systems.

Several types of extraterrestrial phenomena have been discussed by scientists in the context of climate change and mass extinctions of biological species. The first of these is the variations of the Earth's orbit around the Sun. Astronomers have long understood that the direction of the Earth's spin axis in space slowly varies on a time scale of 40,000 years, as does the shape of the Earth's elliptical orbit around the Sun on a time scale of 100,000 years. These well-understood variations in the Earth's orbit do not alter significantly the total solar radiation reaching the Earth's surface; however, they do change the amount of heat reaching different latitudes at different times of the year, thus altering climate. It now seems certain that the variations in the Earth's orbit explain the well-determined climatic changes of ice ages and glaciations over the past several hundred thousand years.

On a much longer time scale, variations in the output of the Sun can be invoked. The brightness of the early Sun was believed to be just 75 percent of its present value. But, as already noted, the carbon dioxide and methane in the early atmosphere of the Earth would have sustained a strong greenhouse effect to keep global temperatures relatively high. Although this greenhouse effect moderated (as carbon dioxide and methane were lost from the evolving atmosphere), the brightening of the Sun maintained the global temperatures. As the Sun rotates around the center of the Galaxy, it will encounter interstellar gas and dust of varying density. If (when?) the Sun meanders through a dense cloud of interstellar gas, it will accrete some of this material and its brightness could increase significantly as a consequence, driving an increase in the Earth's temperature.

Giant solar flares have been suggested as a possible cause of severe disruptions on the Earth. Flares some 100 to 1,000 times larger than any ever observed would be needed to cause major climate changes or the mass extinctions of living species. Some physicists believe that such giant flares could be produced by our Sun (although extremely rarely)

One form of extraterrestrial event that definitely does occur is a massive explosion of a star at the end of its life, a supernova event. Supernovae are real enough; astronomers observe them regularly in distant galaxies. There is no prospect of any star in close proximity to the solar system undergoing such a catastrophic outburst on any foreseeable time scale. Nevertheless, it has been estimated that supernovae occur within the Milky Way Galaxy about every 30 years on average (although they are likely to be hidden from view by dust), and occur close enough to affect the Earth about every 200 million years. It is extremely likely that the Earth was severely affected by a supernova outburst several times during its evolution. Hence, it has been speculated that one or more of the mass extinctions of species might have been precipitated by a supernova. Both the initial radiation burst from a supernova and the particles from the explosion that would reach the Earth several thousand years later would be expected to deplete seriously the Earth's protective ozone layer. Absorption of solar ultraviolet radiation by the ozone layer is an important mechanism for heating the stratosphere, and the serious reduction in ozone levels, even if the ozone layer was to recover in just a few decades, would reduce the temperature at the Earth's surface. Removal of the protective ozone layer could also affect living organisms, exposing them to the full blast of harmful ultraviolet radiation from the Sun.

The Earth's atmosphere provides useful protection against impacts of small pieces of debris left over from the formation of the Solar System; they are burnt up in the atmosphere, and may be seen as *meteors* (popularly known as *shooting stars*). Sometimes many meteors are seen to radiate from a common point in the sky; these are called *meteor showers*. Meteor showers are associated with comets. As already noted, comets come from the cold outer extremities of the Solar System. They are giant accumulations of dust, ice, and organic molecules; the description of a comet being a "dirty snowball" is an apt one, albeit that the dirty snowball is the size of Mount Everest! Occasionally comets are deflected from their freezing hibernation in the outer extremities of the Solar System, and are swept in a highly elliptical orbit toward the Sun. Close to the Sun, some of the ice, dust, and organic material evaporate off the "snowball," being swept backward into the familiar cometary tail by the effect of the solar wind. The tails of comets thus always point away from the Sun.

More massive pieces of debris than meteors can survive the fiery descent through the atmosphere to impact the Earth's surface as *meteorites*. There are many occasions in history when the impacts of meteorites have been recorded. A 1-ton meteorite fell on Nebraska in 1948. The largest meteorite found in the United States weighed 15 tons, and was found in 1902 in Oregon. In 1894 the arctic explorer Commodore R. E. Peary discovered three massive meteorites in Greenland, the largest weighing 34 tons. A great Siberian event of 1908 flattened a forest near the Tuguska River. An impact was recorded by seismographs throughout Europe. Some 100 depressions were found at the site, although not a single meteorite has been recovered. Perhaps the event was the impact of a small comet, since the matter making up the comet would have almost completely evaporated on impact. Its descent to Earth was witnessed by many people; one observer described its passage across the sky as being "like a piece broken off the Sun."

Meteorites tend to contain mostly iron and silicate rock. This suggests that since the meteorites are believed to indicate the composition of the solid material in the inner part of the early Solar System, the nascent Earth is likely to have been a cooling globe of iron and silicates. Because the Earth is an active planet, with a weather system, the evidence of the impact of ancient meteorites has been largely destroyed. Hence, we do not see many craters on Earth as we do on the Moon and planets such as Mercury. But some really large craters resulting from the collisions of giant objects have withstood the obliterating effects of weather and planetary evolution. A spectacular example is the Barringer Crater in Arizona. As noted in the Prologue, meteorites occasionally may be spalled debris from other planets, in the way that the rock labeled ALH 84001 and a meteorite recently discovered in the Sahara desert were shown to have come from Mars.

The extraterrestrial event most favored by astronomers for causing catastrophes on Earth is the impact of comets or meteorites. There have been five mass extinctions evidenced in the fossil records during which it is believed that the majority of the Earth's life-forms were lost. The most recent mass extinction, the so-called Cretaceous-Tertiary (K-T) extinction, occurred some 65 million years ago; victims included the dinosaurs and a host of marine species. There is evidence that the K-T extinction resulted from a giant meteorite colliding with the Earth. Nobel laureate Luis Alvarez proposed this in 1980. The evidence was a thin clay layer containing high levels of the element iridium, at a depth commensurate with a deposit 65 million years ago. Iridium is uncommon in Earth material, but is comparatively abundant in meteorites. The Chicxulub Crater in the Yucatan Peninsula, Mexico, is believed to be the physical scar remaining from the giant meteorite impact that killed the dinosaurs.

In the greatest mass extinction some 240 million years ago, an estimated 90 percent of species were lost. There is some tantalizing evidence that this particular mass extinction may have been caused by a nearby supernova.

What would the impact of a comet or giant meteorite of, say, 10 kilometers diameter, be? The likelihood of an object of this size hitting the Earth has been estimated as about once every 10 to 100 million years. Scientists speculate that in the seconds following the impact of an object of such size, a massive blast wave, similar to, but very much larger than that from a nuclear explosion, would destroy instantly everything within a distance of several hundred kilometers. The intense heat and wind generated in the impact would produce an inferno, which could engulf the globe. Most surface creatures would be fried alive. Only burrowers, cave and nook dwellers, and subterranean creatures would be safe initially. Earthquakes would rock the region of impact for days, and massive tidal waves would destroy coastal habitats. Dust and smoke rising to high altitudes would block out the Sun and precipitate an era of darkness and cold. The plants that survive would be watered by acid rain, and any creatures that survive would breathe a toxic smog. (A few years ago the massive explosions from fragments of Comet Shoemaker-Levy colliding with Jupiter were observed, presenting an awesome specter of what the destructive power of a collision of a comet with Earth might be. It has been suggested that any solar system with an Earth-like planet able to sustain life also needs a Jupiter-sized planet to act as a gravitational attractor for a majority of rogue comets and asteroids, so as to minimize the frequency of catastrophic collisions with the life-carrying planet.)

Meteorite collisions have caught the imagination of science-fiction writers and the makers of Hollywood "blockbusters," presenting intriguing challenges for movie animation and computer graphic experts. Several recent science-fiction movies are based on meteorite or comet collisions.

It is frighteningly plausible that another event of Armageddon proportions could occur. Fortunately, the likelihood of a large object hitting the Earth is extremely small. A 1-kilometer-diameter comet (weighing a billion tons) might hit the planet on average every million years. The energy released in such an impact would exceed by at least a factor of ten the combined explosive power of the nuclear arsenals of the superpowers at the height of the cold war. Comet Hale-Bopp, which thrilled us all in 1996, is estimated to weigh 10 trillion tons. The collision of a comet of such size would be very rare indeed. The probability of collisions is low, but the consequences for human life would be truly devastating.

If a threat were imminent, we now have the means of trying to divert an approaching object from a collision path by launching a rocket with a nuclear payload at it. (A 1-kilometer-diameter asteroid will pass within a few hundred thousand kilometers of the Earth in the year 2028, a "near miss" in astronomical terms.) Some of the technologies developed under the Strategic Defense Initiative ("Star Wars") might be applied to the task. An international organization called the Spaceguard Foundation has been formed to promote and coordinate activities for the discovery of near-Earth objects (meteorites and comets). A network of telescopes will form the Spaceguard System to allow early detection of any object on a collision course with Earth. A 10-year Spaceguard program hopes to detect 90 percent of the estimated 2,000 near-Earth objects with a diameter larger than 1 kilometer that are believed to need careful tracking. Collisions with smaller objects could certainly have grave consequences in the region of the impact, but would not precipitate mass extinctions.

If a catastrophic event meant the end of the human race, which species would survive and take over the Earth? Rodents and cockroaches look like a safe bet, as only human action has restrained their uncontrolled proliferation.

Despite the remaining uncertainties, there is no doubt that science must now consider seriously the suggestions that the climatic and biological histories of our planet should include allowance for extraterrestrial impacts. ETI will have evolved in equally hostile cosmic environments, and would have had to survive mass extinction events.

Observation

From all we know about the present state of our planet and the geological record, we can summarize with some certainty its life history. It was born some 4.6 billion years ago from the disc of dust surrounding the nascent Sun. Meteorites bombarded its surface. Heat generated by the bombardment, and radioactive decay of heavy elements, produced a molten core. Iron, as a heavy element, sank to the center, while lighter materials, such as silicates, created an outer layer of partially molten rock. The liquid surface belched out carbon dioxide, nitrogen, and water vapor, forming the early atmosphere. As the Earth started to cool some 4 billion years ago, it formed a fragile outer crust. Bombarding meteorites fractured the crust so that magma surged out to form massive sheets. The lava from a myriad of volcanoes formed the early continental crust floating on the mantle below. It is possible that a myriad of comets brought water to the planet. Any stirrings of life at this time would have been sterilized by the intense heat produced by the bombardment of planetesimals. From about 3.8 billion years ago, the cooling atmosphere formed clouds, rain poured down to form the early oceans, and lightning filled the sky. A dim red Sun produced only about 75 percent of its present radiation, but an intense greenhouse effect kept the planet hot. A choking atmosphere of carbon dioxide, methane, and hydrogen sulfide would have been poisonous to any modern multicellular organism, but this was nevertheless the crucible of life on Earth. The sky would have appeared pink and the oceans brown. The first life, possibly formed deep in the oceans around hot vents, then moved to the surface. Simple organisms like algae trapped solar energy in the process of photosynthesis, taking in carbon dioxide and releasing oxygen. By some 2 billion years ago, the oxygen had built up to become a major component of the atmosphere. The crust had hardened to become true drifting continents. Some of the oxygen in the atmosphere became ozone, able to filter out harmful solar ultraviolet radiation. Life on land became possible. Oxygen turned the sky to blue, and the sea to blue-green.

Our study of planet Earth suggests that a living planet, one that is able to bring forth life-forms that could evolve to ETI, will have an atmosphere (including a protective ozone layer) and oceans, and is likely to have seasons, a dynamic climate system, a moon (or moons) driving tides, a magnetosphere providing protection from the stellar wind from its parent star, and an active interior. ETI will have survived both planetary upheaval and the impact of extraterrestrial impacts, just as life on Earth has had to do. But there is no reason to suppose that planets around other stars are any less likely than Earth was to develop atmospheres, oceans, climate systems, and life-forms able to find an evolutionary path through the vagaries of planetary and cosmic chaos.

An understanding of our own planet provides plenty of encouragement for the search for ETI.

5. The Emergence of Life

"How often have I said to you that when you have eliminated the impossible, whatever remains, however improbable, must be the truth."

Conan Doyle

We cannot be certain how likely the emergence of life-forms might be on other worlds. We do know that matter appears to have a tendency to form more and more complex structures. Nevertheless, the evolution of even the simplest creatures on Earth appears to embody a random sequence of events so unlikely that even in a vast cosmos the evolution of complex life may be very rare indeed. Certainly it is quite impossible to conceive that the concatenation of peculiar circumstances that led to humans could possibly produce ETI of even vaguely similar anatomy or mentality as humans. Advanced life on other planets will not look anything like humans, so forget about what the movies depict. But the underlying chemistry and physics are the same throughout the cosmos. Hence, there must be every chance that dramatically different concatenations of peculiar circumstances will lead to other forms of complex species, with other forms of intelligence, developing other forms of technology, but all controlled by the same fundamental forces of nature. Science provides a form of conviction on these points, and thus offers confidence, to those prepared to follow the arguments, that SETI will eventually be successful.

Life is based on molecules. This chapter will look at how molecules link together to form the templates for life, and how these molecules replicate (form copies of themselves). Once we understand the molecular basis of life, it is possible to speculate on how life might have started on planet Earth. We can look for evidence in the fossil record. The first simple life-forms were subjected to a hostile environment, so that only those best suited to change survived and flourished. Natural selection has been an unforgiving judge on the course of evolution. The emergence of intelligence in complex life might be understood by studying the best example we have—ourselves.

Organic molecules link together atoms of carbon, nitrogen, oxygen, and hydrogen as their principal components. Carbon shows greater versatility in forming molecules than any other element. A simple example is methane, which has a tetrahedron shape with a carbon atom at the center and a hydrogen atom at each corner. Carbon bonds most readily with oxygen and nitrogen. Many carbon-based molecules contain tens or even hundreds of thousands of atoms, with carbon spread along long chains and in rings. When large molecules requiring vast amounts of information are needed, then carbon provides the perfect spine.

Molecular Genetics

Since the discovery of the microscope, scientists have been studying cells from animals and plants with increasing interest. The cellular division of organic matter was first recognized, and given the name *cells*, by the English scientist Robert Hooke in 1663. It was soon realized that to understand life, scientists had to understand cells, their structure, and their functionality. With the microscopes available by the nineteenth century, the basic structure of the cell was becoming reasonably well understood. Although cells were found to vary enormously in shape, almost all were found to have three basic components: a central *nucleus* (which controlled the cell's behavior), *cytoplasm* (containing various organelles, just discernible with optical microscopes) surrounding the nucleus, and a cell *membrane* surrounding the whole cell. An important group of organelles are the *ribosomes*, because *proteins*, which control all important life functions, are formed in ribosomes. Proteins are naturally occurring complex molecules that are made up of strings or globules of what are called *amino acids* (containing nitrogen, hydrogen, carbon, oxygen, and sometimes sulfur). Another important group of organelles are the *mitochondria*, in which cells convert food to energy. The mitochondria provide the powerhouse for the cell.

A living organism is characterized by its ability to reproduce, and this occurs at the cellular level. Cells reproduce by dividing themselves, a process called *mitosis*. The nucleus of the cell divides first, and then the whole cell divides, with the two halves each taking one of the new controlling nuclei.

Heredity (the passing of characteristics to offspring) was recognized as being a cellular process resulting from the fusion of two cells, an egg cell contributed by the mother and a sperm cell by the father. All inherited characteristics, it was agreed, had to have been transmitted by these reproductive cells.

A fascinating feature of cells was recognized in early microscopic observations. When a cell was resting between divisions, its central nucleus seemed to have a faintly discernible netted appearance. Scientists used certain chemicals to stain their specimens to improve visibility. The netted substance in the cell's nucleus seemed to soak up the stain, and hence was named *chromatin*, from the Greek word *chroma* meaning "color." When the cell suddenly prepared to divide, the chromatin became clotted into small threads, which the scientists labeled *chromosomes*. As a cell divided, the chromosomes were split in two lengthwise, with half of each chromosome going to one of the two new cells where they reemerged into the familiar chromatin netted pattern.

The number of chromosomes in the cells of mature adults was always the same. But here there appeared to be a dilemma. If in the process of procreation a male cell with its full complement of chromosomes fused with a female cell also bearing the correct number of chromosomes, then the offspring of the union would have *twice* the normal number of chromosomes, doubling in each subsequent generation. Clearly, this was nonsense. But nature had learned how to deal with this problem through a chromosome-reducing process in reproductive cells. When scientists recognized the phenomenon, they called it *meiosis*. In a reproductive cell, the chromatin thickened and chromosomes appeared in the normal way prior to mitosis. But there was an important difference. The reproductive cell did split, but the chromosomes did not, contrary to normal mitosis. Instead, each of the chromosomes sought out a partner from among the other chromosomes in the nucleus, and as the reproductive cell broke in two, only one chromosome from each pair was pulled into the newly formed reproductive cell. Each reproductive cell then had only half as many chromosomes as the cell from which it was formed. Thus, when it was later fused with another reproductive cell in the process of procreation, the offspring would have the correct number of chromosomes in its cell makeup.

A normal human was found to have 46 chromosomes (23 from each parent). If just 46 chromosomes could transmit all the complex characteristics of humans, then it was obvious that each chromosome must be responsible for a host of characteristics. By 1915, it was recognized that chromosomes were not continuous thread-like structures, but rather resembled a string of beads. These beads, up to about 1,250 per chromosome, were called *genes*. The science of heredity would henceforth become the science of genetics, the science of the beads.

Let us take a gene that determines a certain physical characteristic, such as height. (When one says that a gene "determines" a characteristic, it is important to remember that this means "all other things being equal." For instance, height in humans has increased dramatically over the past three centuries. This is not a consequence of genetics, but instead reflects improvements in nutrition.) One gene determining height would be received from the male parent and another from the female parent. Both these inherited genes could transmit tallness (in which case the progeny would be tall) or both could transmit shortness (in which case the progeny would be short). In the case of the genes being different, the tallness gene dominates—the "tall" gene appeared to overwhelm the "short" gene. A more influential gene is said to be *dominant*; the weaker gene is said to be *recessive*.

Mathematically there are about 8 million ways the 23 chromosomes from a mother and the 23 chromosomes from a father can combine. The odds that two children of the same mother and father will have the same complement of chromosomes is about 1 in 70 billion. (The exception is identical twins, both of which come from the same fertilized egg and have the same complement of genes.) Since each chromosome has up to 1,250 genes, the probability against two identical individuals being born via the standard selection processes of genetics is a number so large as to be beyond comprehension (it is the number 1, followed by more than 9,000 zeros!). We have taken some time to explain the basis of genetics, as it is at the heart of the diversity of life on Earth, and since genetics is governed by basic chemistry, it will provide the diversity of life-forms on other worlds.

The Magic of DNA

By the 1930s, scientists were starting to probe the chemical nature of the gene. At first, it seemed to be quite simple. Chemical analysis of chromosomes had detected the presence of proteins. Present in many forms in all living matter, proteins direct and speed up life processes, form structural material (for example, bone, skin, tendons, and ligaments), and provide energy for cellular processes. (The proteins that direct and speed up life processes are referred to as *enzymes*. An enzyme is the biological term for a "catalyst," the general term for any chemical compound that speeds up a chemical reaction while not actually being changed by the chemical reaction itself.) Without the critical enzymes, some biochemical reactions could not proceed fast enough to sustain life. In living systems, simple chemical reactions need a helping hand.

Surely the secrets of the genes lay with the proteins? It was true that as early as 1869 a substance called *nucleic acid* had also been detected in chromosomes, but this was thought to be of little consequence compared with the intriguing proteins. In a series of experiments conducted between 1941 and 1944, U.S. scientist Oswald Avery and his coworkers overturned this conservative wisdom. Oswald was following up some earlier work by English biologist Frederick Griffith, in which he had injected into mice two strains of the bacterium that causes pneumonia. One form, called "R," was live but harmless. The virulent strain

"S" had been killed by heat prior to injection. Neither form should have harmed the mice, but some mice died. Postmortem examination revealed the presence of the living form of S. No one had been able to explain this 13-year old mystery until Avery revisited it. He assumed that since the virulent S form was dead, the fatal doses had to have come from the harmless R form that must have changed to the virulent S form through genetic reproduction. Avery and his collaborators cultivated vast quantities of the bacterium to isolate what it was that was converting the harmless R form to the virulent S form. They discovered that it was the overlooked nuclei acid, specifically, *deoxyribonucleic acid*, or *DNA* for short.

The evidence that the behavior of genes is determined by DNA did not initially turn the world of biology on its head. The advocates of proteins as the driving force of heredity were not prepared to give up without a fight. Conclusive evidence was produced by Alfred Hershey and Martha Chase of the Carnegie Institute, Long Island. They took a simple form of virus, called *phage*, that infects bacteria. Phage attacks a bacterial cell by using the cell's chemical apparatus plus its own genetic information to mass-produce new phages. Phage was known to be made up solely of DNA and protein. Hershey and Chase separately labeled the protein and DNA with different types of radioactive material that would allow their subsequent passage to be traced. By this means they were able to show that only the DNA part of the phage actually penetrated the bacteria's cell wall. Therefore, it must be the DNA alone that was affecting the bacteria's genetic behavior. Now no one could doubt that DNA was the driving force of heredity.

Even before Hershey and Chase produced the conclusive evidence of the determining role of DNA in genetics, chemists had made great strides in determining DNA's chemical makeup. DNA is made up of three types of ingredients: simple sugars of a type called *deoxyribose*, phosphate units, and finally nitrate compounds. The nitrate compounds, known as *bases*, were of four kinds: adenine, thymine, cytosine, and guanine, abbreviated as A, T, C, and G.

Knowing the chemical composition was a long way from

knowing how these various components were linked together. And how did DNA work? X-rays played a key role in bringing the genetics revolution to a triumphant conclusion. A technique called *X-ray crystallography* had been developed and was used to determine the internal structure of complex materials. X-rays, like all forms of waves, undergo a phenomenon called *diffraction*. Diffraction is essentially the ability of a wave to creep around the edge of any obstacle it encounters—think of how a sea wave wraps around a jetty. An interesting demonstration of this effect can be given for visible light using a diffraction grating, a clear screen etched with thousands of microscopic grooves. Each wavelength of an incident beam of white light bends around the edges of the grooves to a different extent, hence, splitting the white light into a spectrum of its component colors. Thus, a white lamp viewed through a diffraction grating would appear as a stretched-out series of overlapping images of the lamp in the different colors of the rainbow. Diffraction gratings act with the light being either transmitted through the grating or reflected from its surface. (Hence, you can see a similar rainbow effect in light reflected off the surface of a compact disc, which is engraved with "flats" and "pits" in a circular pattern.) Diffraction of X-rays has been shown to be a useful method for establishing the three-dimensional structure of matter. Instead of using grooves in a diffraction grating to produce a diffraction pattern, the X-ray crystallographers depend on the regular array of atoms and molecules in certain materials to produce a similar effect. They then have the challenge to interpret the complex pattern of illumination diffracted by crystals of the material to give the three-dimensional structure of the intervening atoms and molecules responsible for producing the diffraction effects. The task of interpretation is exceedingly complex; nevertheless, X-ray crystallography has proved to be a technique of immense value. It enabled the complex structure of DNA to be determined.

Unraveling the structure of DNA was one of the heroic achievements of science. The breakthrough was made in 1953 by James Watson and Francis Crick at the Cavendish Laboratory in Cambridge, England. Their famous double-helix model for

DNA was based on X-ray crystallography data obtained by Maurice Wilkins of Kings College London and Rosalind Franklin (who sadly died before her achievements could receive due recognition).

Crick and Watson showed that DNA is made up of two intertwined helical strands, connected by the A, T, C, and G bases. The bases are paired only in certain ways, so that A always twins with T, and C always twins with G. A strand of DNA is then an apparently random order of base pairs, say AT, GC, CG; TA, TA, CG; AT, GC, and so on. A single gene is then a strip of DNA containing perhaps thousands of base pairs. The bases A, T, C, and G constitute the "alphabet" of the genetic code, and triplets of bases form the "words." The DNA in a human cell contains billions of base pairs. Clearly in such a complex genetic code, there is scope for error—a CG where a TA should be. Hence, there is the chance that genetic malfunctions will occur, especially since genes can be mutated (damaged) by external sources of radiation, such as radioactivity. However, understanding the genetic makeup of people holds out the hope of correcting mutations, and compensating for the worst effects of inherited defects. And over the eons of evolutionary history, mutations have led to the emergence of new living species.

As already noted, a clear requirement of any genetic material is that it must be able to reproduce itself. This capability is neatly built into the double helix. When chromosomes duplicate, the whole DNA structure "unzips" down the middle. All the A bases separate from T bases, and the G bases separate from C bases. Then each of these half strands grows its other side, uniquely replicating the original since all T bases are constrained to link with A bases, and all G bases are constrained to link with C bases.

The genetic code contained within DNA controls life through the production of proteins. DNA remains locked within the nucleus of a cell, but it directs precise instructions to other parts of the cell. It does this by unlocking short strands of what is called *messenger RNA* (ribonucleic acid). In RNA, uracil (U) substitutes for T; the other three bases and the linking rules remain the same as for DNA (with U now linking to A).

Messenger RNA carries the blueprint for protein production from the nucleus to the ribosomes where the proteins are manufactured. DNA also produces another form of RNA, called *transfer RNA*. Transfer RNA seeks out the required amino acid building blocks of the proteins, and these are then locked into the required sequence by the messenger RNA. For example, the triplet CUU in RNA instructs a cell to add the amino acid leucine to a growing strand of protein. It is quite remarkable how DNA produces precise directions for the production of proteins in the correct quantities and at the moments they are required. Once the magical properties of DNA were understood, then an inherent logic in genetics was revealed.

The discovery of the double-helix structure for DNA, and its genetic consequences, had a monumental impact on molecular biology, and allowed scientists to speculate on the origin and evolution of life with a new authority. All species that are or ever have been on Earth since the emergence of the first simple DNA-based organism were formed from an assemblage of amino acids that make up DNA, the sequence of the base pairs determining their individual characteristics and behavior. For life, DNA is "it." DNA allows life to reproduce; it allows for variations in heritable characteristics so that descendants can develop new traits; it enables evolution to bring forth diverse species. The discovery of DNA unlocked the secrets of life, and therefore stands as one of the greatest of all scientific breakthroughs. Understanding how DNA shaped life on Earth provides us with the scientific tool necessary to contemplate the emergence of life elsewhere in the cosmos.

The Fossil Record

In recent times, the tracking of DNA has provided the greatest insight to evolution. In earlier times scientists looked to fossils to explore the past; the study of ancient fossils offered a way of understanding the origin of life on Earth, and its meandering path to intelligence.

Fossils are the remains of life-forms of previous epochs preserved in sediments, volcanic ash, coal, and oil. Remains can also be preserved by being trapped in amber, frozen in ice, or mummified in arid or rarefied conditions. The study of ancient life, in its geological context, is called *paleontology*, the study of bones and stones.

The oldest fossils of undisputed age are 3.8 billion-year-old cyanobacteria (blue-green algae) found in Australia and South Africa. Most fossils showing a diversity of life-forms are very much younger, typically less than 500 million years old.

The challenge in studying ancient life and ancient geology is to try to put it in a context where the enormous time scales (of millions and billions of years) can be understood against the more familiar times of human experience (a few tens of years of personal experience, or a few thousand years of recorded history). A familiar illustration is to picture the history of the Earth as a single 24-hour period, commencing at midnight. The earliest fossils depict life at dawn. By noon, when half of geological time had passed, there was sufficient oxygen for more complex life-forms to start to evolve, although life remained relatively simple. Sometime after 9 P.M. there was an explosive development of new complex life-forms. The dinosaurs appeared at 10:42 P.M. Humans appeared about a minute and a half before midnight. Civilization is depicted by a single second! Television, computers, and nuclear weapons have only been around for a few milliseconds on this compressed time scale. We really are merely a blink of the eye in the evolutionary day.

The time periods within which fossils are dated are defined by the ages of rocks, which can be measured as far back as 4 billion years. The method for dating rocks uses different forms of the same element. These different forms are called *isotopes*. The technique for dating is based on the comparison of a stable isotope of a certain element with its radioactive isotope (where this exists). The time it takes for half a radioactive isotope to decay away is called its *half-life*. Thus, for example, radioactive uranium decays to lead with a half-life of 4.5 billion years. (Lead is referred to as the "daughter" product of uranium.)

By dating the deposits in which the fossils are found, the time history of the fossil record can be established. The major periods of paleontology are referred to as *Paleozoic* ("ancient life"), *Mesozoic* ("middle life"), and *Cenozoic* ("recent life"). These major periods are then further subdivided, with periods often called after the region where fossils have been found. For example, the Jurassic period is named after the Jura Mountains in France and Switzerland, and the Cambrian period after the Latin name for Wales. (The Cambrian period, which started some 530 million years ago, saw the explosive evolution of diverse complex life-forms.)

The problem with paleontology is that the fossil record gives only a very partial picture of life's history. The history of the Earth's crust has been too violent to preserve other than a small random sample of fossils. The bones of vertebrates are preserved preferentially. Rarely are conditions right to preserve the remains of soft-bodied organisms, which succumb to scavengers and bacteria. Although several hundred thousand fossils have been classified and named, this is thought to represent less that 0.001 percent of all the species that ever existed. The fossil record is scant indeed, but nevertheless has provided great insight into the complexity of evolution. Paleontologists have faced a daunting task writing the history of life, a task akin to writing poetic prose while limited to only five letters of the alphabet.

Natural Selection

It is believed that all life-forms on Earth are interrelated, with a complex pathway leading back to the very first life-form some 4 billion years ago. There is a "common ancestor" back there somewhere. From this common ancestor, possibly originating at a deep-sea vent in the primordial oceans, life in all its complexity is believed to have evolved down many pathways and roads. Yet if the whole of the living world came from a single common ancestor some 4 billion years ago, how did life diversify so spectacularly to produce the millions of current species, let alone the billions that we know have died off in the past?

The single scientific work that revolutionized the understanding of evolution was Charles Darwin's *On the Origin of the*

Species by Means of Natural Selection, first published in 1859. Contrary to what is often said, the majority of scientists already accepted evolution as a fact before Darwin. However, it was Darwin who first offered the correct explanation of this historical process and revealed its ubiquity, at the same time drawing it to the attention of the public. Darwin drew on the ideas of many others, as well as his own extensive fieldwork. He had withheld his theories for almost 30 years, conscious of their likely impact on the sensitivities of Victorian culture. The scientific revolution he launched was more fundamental and profound in its impact than he could ever have envisaged.

The key elements of natural selection are as follows. All organisms produce more progeny than can hope to survive or reproduce. Characteristics of parents appear in their progeny, but with the possibility of mutations introducing variation into the species. The organisms that do survive are then those best adapted to their local environment, which will change with time (perhaps spectacularly so, in even relatively short time intervals). Generation by generation, thousands and millions of times over, the organisms best adapted to a changing environment will survive to pass on their characteristics to progeny, giving them the advantages to survive. This was "survival of the fittest."

An understanding of the nature of evolution provides us with a useful definition of *life*: Life is a self-sustaining chemical system capable of reproducing itself and of undergoing Darwinian evolution.

The fossil record suggests that natural selection has been, highly spasmodic process, rather than a process of ponderous but continuous change. There have been long phases over hundreds of millions of years with little apparent change, followed by intervals of rampant development and loss. Development booms may relate to the conquest of new richer and safer environments, while "busts" may have been precipitated by terrestrial (for example, earthquakes, volcanoes, tidal waves, fires) or extraterrestrial (for example, cometary collisions, supernovae) catastrophes. "Bust" for one species could prove to be "boom" for others, as the demise of their natural predators became part of evolutionary history. An intriguing feature of the evolution of life has been the stability of its bacterial mode from the beginning of the fossil record until today. Classes of simple microbes have shown themselves to be spectacularly robust to events that proved to be calamitous for more complex species. More heat, less heat, more water, less water, higher pressure, lower pressure, volcanoes, floods, earthquakes, comet collisions—there have been bugs that survived it all. Life on other worlds in simple form will have proved itself equally robust to the chaos of a violent universe; hence, our contention that "slime" in the cosmos will prove to be commonplace, even if ETI is probably rare.

Evolutionary Evidence from DNA

Since the time of Darwin's classic work, scientists have sought to establish the relationships between different living species and to track their common ancestry. The standard way of doing this was to look for common characteristics in the anatomy of different species. Thus, for example, birds would be classified as one inclusive group, within which different types were grouped based on certain anatomical characteristics. This method of classification is called *cladistics*. Its development over the years as a research tool has produced various possible evolutionary paths for the principal species. Humans are seen as being part of the primate line, of which the African and Asian apes are also offspring.

The ability in recent years to analyze the DNA and proteins of living things has provided scientists with a more precise method of defining the evolutionary relationships between species. The more similar the DNA and protein sequences from two species are, the more closely related they are. We refer to them as having a recent "latest common ancestor." (Of course, all species are believed to have had the same earliest common ancestor. The "latest common ancestor" refers to the evolutionary "crossroads," where the species being compared parted company from each other.) The common ancestor of two species with very distinct DNA sequences lived much longer ago than the latest common ancestor of two species with very similar sequences.

Humans share 98.4 percent of their DNA with chimpanzees, and 97.7 percent with gorillas. This does not mean that humans evolved from chimpanzees; it does mean that humans and chimpanzees followed separate evolutionary paths from a latest common ancestor. African apes are more closely related to humans than are Asian apes. The Asian apes parted company from the human evolutionary path some 10 million years ago, whereas African apes and humans took different paths along the primate line about 5 million years ago. Modern humans appeared just 100,000 years ago, with the physical characteristics we have and inquiring minds.

Simple Cells

It is possible to define three fundamental divisions of living things. The first we call *archaebacteria* (ancient bacteria), the most primitive form of bacteria that only survives today in extreme conditions, such as in hot pools at approximately 55 degrees centigrade. Such bacteria die instantly when exposed to oxygen. They are the remaining relics of life in the primordial Earth, before the formation of an oxygen atmosphere. They have survived only by hiding in oxygen-free environments. (Since archaebacteria survive only at extreme temperatures, it has been inferred that high temperatures were needed for the origin of life, although some scientists have challenged this inference.) The second fundamental division is the *eubacteria*, the many thousands of species that are important recyclers of many types of vital molecules. Without the diligent work of the eubacteria, other life-forms could not exist. These two divisions go under the collective name *prokaryotes* ("pre-seeds"), so-called because they carry their DNA lose within the cell membrane; they do not have a nucleus. The third fundamental division of cells is the *eukaryotes* ("true seeds"), which package their genes in the nucleus of a cell. (Excluded here are the "viruses." Despite having fragments of DNA or RNA, they cannot survive in isolation. They can only replicate by invading and subverting living cells.)

We now recognize that various forms of prebiotic molecules, such as carbohydrates, amino acids (the building blocks of proteins), and nucleotides (the building blocks of DNA and RNA), can be produced when there is plenty of energy (for example, as on the primitive Earth from lightning, ultraviolet radiation, and meteorite collisions) and little free oxygen. Experiments simulating early Earth conditions were performed by Stanley Miller at the University of Chicago in the 1950s. A flask of water (simulating the ocean) was heated, forcing water vapor into a primitive "atmosphere" of methane, ammonia, and hydrogen. The atmosphere was then exposed to a continuous electrical discharge simulating lightning. The experiment yielded many amino acids and other interesting organic compounds, which could then "rain" into the simulated "primitive ocean." It seemed that creating the organic building blocks of life would be a straightforward process on the primitive Earth; all that was needed was an ocean, an atmosphere, and lightning. A problem was that Miller's primitive atmosphere needed ammonia and methane, and other scientists were arguing that the primitive atmosphere was dominated by carbon dioxide and nitrogen. Experiments simulating this type of atmosphere produced organic molecules, but a less rich variety than what was produced by Miller's experiments.

An alternative theory is that the amino acids arrived from outer space, brought by the asteroids and comets bombarding the young planet. Organic molecules are common in the cosmos; for example, the spectral signature of ethyl alcohol (the drinkable form of alcohol) and formaldehyde (the embalming kind) have been detected in interstellar clouds. Perhaps the amino acids were first created in interstellar space, then frozen into the comets that bombarded the nascent Earth. Just how organic precursors such as formaldehyde, ammonia, hydrogen cyanide, and the likes could combine to form the amino acids in the rarefied conditions of the interstellar medium remained uncertain until some recent research was completed. At NASA's Ames Research Center, scientists tried to simulate conditions in a dust cloud in the frozen depths of interstellar space. This required producing conditions in the laboratory more extreme than those attempted by Stanley for the primeval Earth. Temperatures just 10 degrees above the absolute zero for temperature (at −273 degrees centigrade) are needed, and a near-perfect vacuum. Ultraviolet radiation simulating starlight is directed into a mixture of molecules found in the type of interstellar clouds in which stars and planets are born, such molecules as methane, carbon monoxide, and ammonia. The ultraviolet intensity was set so that 1 hour in the laboratory simulated a thousand years in space. On a grid simulating the conditions on frozen dust grains, remarkably complex organic molecules were formed, including amino acids. Trapping the basic organic molecules on the frozen surfaces of dust grains seems to allow some complex chemistry to take place.

It appears to be tantalizingly plausible that the seeds of life formed on dust grains in the cold depths of space, and were then brought to Earth (and a myriad of other planets in the cosmos) via a hail of comets, the frozen junk left over when a star and planetary system are born. With this hypothesis, while still surrounded in massive uncertainties, the creation of life did not require anything special about planet Earth. It merely required the Earth, and any other life-bearing planet in the cosmos, to provide a fertile breeding ground for the seeds of life produced in a distant cosmic nursery. Life does not become a chemical option on this picture; it becomes a cosmic certainty.

Certainly the cometary debris from the formation of the Solar System is known to contain amino acids, and could have brought them to Earth in the initial period of frequent cometary collisions. Yet another theory involves the early Earth passing through an interstellar dust cloud, acquiring copious amounts of organic material that had been created on dust grains.

Progressing from the prebiotic molecules (whatever their origin) to DNA is the challenge. No experiment has ever been able to produce DNA in a test tube (although some intriguing experiments on RNA are being attempted). The step between prebiotic molecules and self-replicating DNA remains a matter of speculation.

By what possible string of chemical reactions could the interdependent systems of DNA, RNA, and proteins have been brought into being? Nucleic acids (DNA and RNA) are presently synthesized only with the help of proteins, and proteins are only synthesized if the nucleic acids provide the blueprint. It is inconceivable that nucleic acids and proteins, both of which have enormously complex structures, arose spontaneously at the same time in the same place. The resolution to this paradox may lie in an "RNA world" that preceded the present evolutionary DNA world, although such a suggestion passes the problem to an earlier phase, and we still have to ask, What created the RNA world?

There is actually very little evidence to support the hypothesis for an RNA world. However, the idea has certain attractions. It has to be imagined that during the routine chemical reactions taking place in the matter in the primordial oceans, bonding atom to atom, a remarkable molecule was formed essentially by chance. (There was time aplenty for some pretty unlikely organic chemistry to have occurred.) This molecule was remarkable because it could create copies of itself; it could "replicate." It might not have been the most complex molecule around, but it only had to happen once. Then, by virtue of its ability to replicate, it would draw on the material making up simpler molecules. It thus makes it less likely for molecules more complex than itself to form, since it made first call on the simpler component parts. This remarkable molecule, which is believed to be closely related to what we now know as RNA, would have to replicate itself without the help of proteins. (Scientists have been able to get RNA to replicate in the laboratory, with help. But this is not yet "life in a test tube," although it is getting close, because the RNA molecules do not yet entirely control their own replication.) Before any simple microorganisms were formed— indeed before DNA and proteins were formed—an "RNA world" could have existed where RNA molecules in the oceans replicated, mutated, and underwent a form of Darwinian evolution. This initial RNA would have had to "invent" proteins (or perhaps RNA and proteins co-evolved in protocells, rather than evolving separately). It is then surmised that DNA evolved from this RNA (although continuing to make use of RNA for its own purposes), and took over from RNA the role of the guardian of heredity. RNA then vacated center stage in the drama of life to make way for DNA, but kept an important supporting role.

Chemists now believe that even RNA might not have been the first replicating molecule. It might have had a simpler replicating precursor, from which RNA itself then evolved (displacing the first replicator, of which there is now no trace), or there might have been a succession of ever-simpler replicating precursors. The precise events leading to an RNA world remain a matter of speculation, and until a plausible theory is developed, the exact nature of the origin of life on Earth remains a mystery. However, attention is now focused on trying to understand how four stages could be linked. The first stage involves the organic molecules widely observed throughout the cosmos, which might have linked to form amino acids on dust grains in interstellar clouds and then were transported to Earth by comets. The second (and most uncertain) stage is a pre-RNA world on Earth where a first simple replicator appeared. The third stage is the hypothetical RNA world itself. And the fourth stage is the DNA world we now live in that brought forth such diverse life-forms. As on Earth, so on planet "X" orbiting solar-type star "Y"? Once we can make contact with ETI, we should be able to determine the truth.

There is a major unresolved problem with the idea that an RNA (and possibly an earlier) world preceded the DNA world. The bases of RNA quickly break down in hot water. At the boiling point of water, all traces of the RNA bases A and G would disappear in a few decades. Thus, it can be concluded that the bases could not have accumulated in the hot parts of the primordial Earth, just those parts (such as deep-sea vents) much favored from other arguments as being the sites of the first replicators. The bases can survive for millions of years at low temperatures, and therefore it has been suggested that ice on the ocean surface might have been the site of the all-important prebiotic chemistry. Certainly there are still some major questions hanging over the concept of an RNA world. The details of how the first replicators emerged might be uncertain, but it happened quickly, so it is perhaps reasonable to assume that it happened easily.

Once the first replicating molecules were in place, a new form of order on the primordial Earth came into being. At some point, and by an unknown process, the replicating molecules found comfort within a simple membrane, forming the first cells. The complex chemistry of RNA and DNA needs to be contained within a boundary; if it were left free circulating in water, the density of the key molecules would not be sufficient for the important chemistry of life to happen. Just how the first cell membrane formed is a mystery. It was a key moment. The earliest cells probably used chemical rather than photochemical energy sources. The simplest organisms, the archaebacteria, began to proliferate. The starting pistol had been fired on the Darwinian evolutionary race of life.

The simple archaebacteria found no difficulty flourishing in the primitive Earth environment, with plenty of energy. The indications are that they appeared very early indeed in the Earth's history. Carbon exists in two stable isotope forms: one containing six neutrons and the other, slightly heavier, containing seven neutrons in the nucleus. The lighter isotope can be absorbed more easily by living matter. Some sedimentary rock in Greenland, estimated to have been laid down 3.8 billion years ago, shows an excess of the lighter carbon isotope, suggesting that it must have contained life. Yet 3.8 billion years ago was only about 200 million years after the Earth had cooled to the point where even archaebacteria could have survived the extreme heat. It seems that life appeared relatively quickly (in evolutionary terms) after the conditions on the Earth's surface made any form of life possible.

Although over a period of some 3 billion years life diversified, it remained in the realm of bacteria, a period that occupied over 80 percent of geological time. The explosive evolution of life awaited the Cambrian period, a little over 550 million years ago. The Precambrian period, occupying the bulk of evolutionary history, was essentially a period of microscopic life. Fascinating relics of this primitive world are massive "rocks" called *stromatolites*. These are a mixture of layered microorganisms and rock. The outside is a spongy coating made of cyanobacterial filaments, from which a sticky mucus that protects them from

environmental contamination has been secreted. Grains of sedi-
ment trapped by the mucus become rock. The filaments creep
sunward, producing a new layer around the stony interior. Thus,
the stromatolites formed over the eons are dead rock in the center and living matter on the outside, an interesting partnership
between the organic and inorganic worlds.

The Emergence of Complex Life-forms

Around 2.5 to 2 billion years ago, the Earth had cooled to the
point where the crust formed land masses. In the shallow waters
off the newly formed continents, primitive photosynthesizing
bacteria evolving in an oxygen-free atmosphere consumed and
gave off other gases. At some point, mutant offspring began to
produce oxygen, poisoning their oxygen-hating parents. Oxygen
haters had to seek refuge in deeper waters. The new oxygen-pro-
ducing bacteria (the *cyanobacteria*, also known as *blue-green
algae*) exploited carbon dioxide, water, sunlight, and time, to
transform the Earth. They converted carbon dioxide into ener-
gy-rich sugar, with oxygen as a waste product.

Photosynthesizing life-forms anchored the Earth's food chain,
providing the sustenance for all higher life-forms. The oxygen
they produced first oxidized the metals such as iron in seawater.
The oxidized iron precipitated as "rust" on the ocean floor. (The
steel we use today for buildings, cars, and bridges is made from
iron ore created by the oxygen from primitive bacteria.) When
all the metals in the ocean had oxidized, oxygen was freed to the
atmosphere, sweeping out of the atmosphere the toxic fumes of
methane and sulfur dioxide. As the air became richer in oxygen,
the oxygen-loving bacteria took over the surface waters, driving
the oxygen haters farther into retreat. A planet more hospitable
to advanced life-forms had been created. The diverse biochemistry of the simple microbes allowed them to survive and flourish
in widely different environments.

The development of eukaryotic cells may have arisen by their
being penetrated by prokaryotic cells (carrying DNA and ribosomes in the cytoplasm). The mitochondria in eukaryotic cells

that burn sugar and oxygen to generate energy may merely be remnants of oxygen-breathing bacteria that invaded eukaryotes some 2 billion years ago. On this interpretation, the reason why we need to breathe oxygen is to feed the descendants of parasitic bacteria in our cells. Similarly, plant cells contain organelles called *chloroplasts* that convert sunlight into chemical energy, and these are believed to be from former cyanobacteria.

The eukaryotes appear to have evolved very slowly at first. The arrival of sexual reproduction sparked more rapid evolution. Prior to about 1.2 billion years ago, organisms reproduced asexually, splitting to produce exact copies. But advanced eukaryotes developed a more novel approach, involving two parents, mixing the genes together to produce new combinations. The discovery of sex forced evolution into overdrive.

Sexual reproduction seems to be just one form of advanced life-form behavior inherited from simple cells. Another of interest is the ingestion of food by a simple amoeba, which we see in the amoeba-like behavior of white blood cells fighting infection.

The Path to Humans

The evolutionary path leading to humans has been the fortuitous outcome of thousands of random events. The nonoccurrence of any single one of these events might well have precluded humans from coming into existence. Intelligent life might never have developed on Earth, or it might have taken a very different form. To demonstrate this point, let us consider just four of the millions of coincidences that ended up with the emergence of us:

1. First coincidence: One of the few survivors of the early-Cambrian evolutionary explosion was a small swimming creature with a dorsal stiffening rod, the ancestor of the vertebrates.

2. Second coincidence: A group of lobe-finned fish evolved bones strong enough to bear their weight, so later when their

descendant vertebrates moved out of the oceans, they were able to lumber across the land.

3. Third coincidence: The collision of an asteroid with the planet 65 million years ago wiped out the dinosaurs that ruled the Earth, allowing mammals to enter the race for domination.

4. Fourth coincidence: Climate change and continental drift 7 to 5 million years ago allowed grasslands to encroach into the dense forests of Africa, forcing prehuman apes out of the trees. Our ancestors did not so much abandon the trees; the trees abandoned them.

Scientists have little idea what triggered the Cambrian evolutionary explosion some 530 million years ago. Quite simply it is the most amazing and puzzling event in the history of the Earth. (It is so puzzling that some scientists suggest that it could never have happened the way the fossil record suggests, that it is merely an artifact of the data with all its inherent uncertainties. They are in a minority. Our description of it here is rather simplistic, because a full understanding requires a much lengthier presentation. Those interested are advised to read Stephen Jay Gould's classic book on evolution and the Cambrian explosion, *Wonderful Life*. It is a fascinating read.) Single-cell organisms had existed in the oceans for 3 billion years, showing increasing diversity but little progress in evolutionary terms, at least in terms of emerging complexity. Then in just 5 million years, a mere blink of the eye in evolutionary terms, the world went mad with biodiversity. In those 5 million years all known forms of modern life emerged except one (the Bryozoa, a class of marine organism that appears to have emerged at the subsequent Ordovician period). Subsequent mass extinctions may have killed off vast numbers of species, but they have always left enough to allow evolution to proceed again, as survivors and their mutant offspring fill up the ecological niches left vacant by those who did not survive. Emerging from the Cambrian, bacteria had oxidized the Earth, single-cell organisms had stumbled on the evolutionary benefits of two-way sex, and animals had moved to a different scale of complexity and size. Darwinian evolution had finally made the big time!

One plausible explanation for what put evolution onto the fast track depends on the greenhouse effect. There is geological evidence that between about 750 million and 570 million years ago, the Earth experienced a protracted period of extreme cold. Several ice ages ran back-to-back. Most simple life existing at the time was killed off by the extreme cold. Then, about 570 million years ago, a period of volcanic activity spewed greenhouse gases into the atmosphere and created a giant thaw. The warming planet created a new environment to boost the evolution of new life-forms.

Over 3 billion years of simplicity, followed by 5 million years of intense creativity, and then 500 million years of an evolutionary lottery led to millions of species, of which we are but one. After the Cambrian explosion that produced an enormous diversification of species, evolution settled subsequently into a stuttering path of decimations. It is wrong to envisage life becoming more and more diverse, and more and more complex, as Darwinian evolution proceeded apace from its simple beginnings. Few of the great diversity of species from the Cambrian explosion survived. This could not have been because they were not suited for their environment, since they clearly appeared and survived in the first place. Some event, or sequence of events, resulted in the loss of most species. A new period of diversification was halted by a further decimation. And so evolution struggled onward, with a period of diversification brought to a shuddering halt by a further interval of decimation. After each decimation, evolution restarted its diversification from the base of the few remaining species. It is tempting to suppose that the decimations resulted from the terrestrial and extraterrestrial catastrophes described at the end of Chapter 4.

A simple picture of all this is the heavy pruning back of a rose bush. The "rose bush" of evolution sprouts many new side shoots of new species, and these in turn produce their own offshoots. So a complex system of new species develops, all of which can be tracked back to the main stem of the evolutionary "rose bush." Then some catastrophic "pruner" comes along (a comet, a supernova, an ice age, or so forth) and cuts back the "rose bush" of life drastically. But the main stem and some established offshoots remain, and strong growth is soon reestablished.

The post-Cambrian evolutionary lottery thus seems to have depended in large measure on random catastrophes. As previously noted, the mass extinction at the time of transition from the Cretaceous to the Tertiary epochs some 65 million years ago, which killed off the dinosaurs, is believed to have resulted from the impact of a 10-kilometer-diameter asteroid. The dinosaurs were killed off, but mammals survived. Mammals and dinosaurs had coexisted for at least 100 million years. But the mammals survived the cosmic catastrophe, possibly because they were smaller and some could hide from the inferno. When the continents were all grouped together as the supercontinent Pangaea, the earliest mammals were able to radiate to all parts of the world, and subsequently evolved to become the largest of beasts (for example, the African elephant and the blue whale).

The first primates appeared in the tropics. They were tree-dwelling vegetarians that developed the characteristics needed for such a habitat. An opposable thumb provided grip, and then skills (which stimulated intelligence). Sharp eyesight (rather than a sensitive nose) was needed for a jungle environment, and forward-looking eyes allowed the sense of perspective needed for swinging from tree to tree. The primates developed larger brains than ground-living mammals.

The hominoid primates consist of the lesser apes (gibbons and siamangs), the great apes (chimpanzees and gorillas), and humans. The ancestors of modern humans represented at least nine hominid species. The direct ancestor of *Homo sapiens* (modern humans) was almost certainly *Homo erectus*, but there are different theories for the path to *H. erectus*, with uncertainties remaining because of the incompleteness of the fossil record.

Homo habilis, prehuman apes, came before *H. erectus*. During a long period of global temperature fall some 7 to 5 million years ago, the tropical forests of Africa succumbed to encroaching grassland. Some hominoid primates were forced from the trees. They stood up to explore a more open environment, and the time that they staggered onto their hind legs and made a few faltering steps was one of the most exciting events of evolution. Since the prehuman apes were short, and the grass was high, it has been suggested that they were forced onto their hind legs just to be able to see their environment.

About 3 million years ago brain sizes increased significantly (by about four times), and bodies developed certain human-like characteristics. *H. habilis*, "the handyman," had strong hands and developed basic tools and simple craft skills. *H. habilis* added meat to the diet, although it is not clear whether they were scavengers or hunters. Tools were used to scrape meat from the carcasses of dead animals. More meat meant even bigger brains, ensuring that more could be learned to improve their chances of survival. What greater motivation to learning is there than to learn how to survive in a hostile environment. As it was then, so it is now.

Two theories dominate current debates about the path from *H. erectus* to *H. sapiens*. The "multiregional" theory suggests that *H. erectus* spread from Africa around the world (to Europe, China, and Australasia), and then began to develop regional (racial) characteristics. The "monogenesis" ("out of Africa") theory suggests that *H. erectus* evolved to *H. sapiens* in Africa, and it was *H. sapiens* who then spread out across the world seeking new hunting grounds and less hostile environments. The "out of Africa" theory is probably winning by a short head.

Tool-making skills were improved, and a major new technology discovered: fire. *H. erectus* became a regular and skilled hunter, now making a significant impact on other species. A hunter needs more territory per person than a vegetarian, and this probably drove the sudden migratory expansion out of Africa.

H. erectus evolved to *H. sapiens*, and in Africa and Asia, *H. sapiens* evolved to modern humans. One branch of *H. sapiens* in Europe survived from about 100,000 to only 35,000 years ago, then died out. This was the Neanderthal species. The Neanderthals developed more sophisticated tools, made clothes, and held ceremonies. There is evidence that they were a sensitive race, with a spiritual awareness. Nevertheless, they did not survive.

H. sapiens, "the wise men," started to use symbols to represent things; carvings and paintings appeared. They became skilled communicators. The stage was finally set for the emergence of the species we now recognize as intelligent humans.

The emergence of intelligence remains poorly understood. What defines intelligence? Certainly, a language is a sign of intelligence. So is versatility; a long life promotes versatility by offering the opportunity to discover new patterns of behavior. A further sign of intelligence is foresight (an ability to plan ahead) and the ability to learn from experience. (Social behavior gives individuals the chance to mimic useful discoveries of others.) It is thought that intelligence emerged through the refinement of some brain specialization, and a good candidate is the ability the human brain developed to put things into sequence.

The ability to put things into sequence is very evident in language. The orderly arrangement of verbal ideas is called *syntax*. Without syntax, we would be no more clever than the chimpanzees. A chimpanzee uses about three-dozen vocalizations to convey about three-dozen meanings. Humans also have about three-dozen vocalizations, but have learned to string these together into words, and to string words together into sentences. The development of syntax is one of the most important developments in the evolutionary transition from apes to humans. How did humans move from the ape state of one sound to one meaning, to the human ability to create sequential combinations of sounds to provide a vast range of meanings? Unfortunately "language leaves no bones," so we are lacking for language the equivalent of the fossil and DNA evidence that enables us to track physiological change.

The human ability to string things together does not just show up in language. It is evident in stringing notes together to form melody, steps together to form dance, or rules together to form procedures. All these abilities are likely to be linked to the brain's ability to plan what is called *ballistic movement*, a sequence of movements that once started is carried to completion. Throwing a ball and hammering a nail are examples of ballistic movements. Such ballistic movements need the brain to schedule the exact sequencing of dozens of muscles. Ballistic movements would have been favored by Darwinian evolution, allowing the effective development and use of hunting tools and weapons, important additions to hominids' basic survival strategies. Thus, our love of rhythm, rhyme, and rules comes from primitive survival instincts. So when the music starts, do not be embarrassed by the desire to boogie; the need is "hardwired."

The 2-millimeter-thick cerebral cortex is the part of the brain most involved in making novel associations. If flattened out, the human cerebral cortex would cover four sheets of paper; a chimpanzee's cortex, one sheet of paper; a monkey's cortex, a postcard; and a rat's cortex, a postage stamp. It is thought that here lies the site of the brain specialization so paramount for intelligence. Mouth movements for the linking together of sounds in language require the same sequencing ability as ballistic hand movements. Since accurate hand movements improved hunting, and hence the chance of survival, the ability to link sounds into language may have come as an evolutionary bonus of our ancestor's survival instincts.

For the first 2 million years of human existence, sources of food depended on the whims of nature. The great thaw at the end of the last ice age produced the spread of grasses that became modern cereal crops. Humans learned to settle the land and control their own food supply; the hunter could settle. The earliest settlements were merely groupings of the first farmers. But as crop surpluses grew, farming villages became organized trading centers. Towns, and eventually cities, required planning, architecture, and legal codes. Humans had become a social species. Modern civilizations started to emerge. Technologies started to be developed. And here we are. We did not ask to be masters of the Earth. But evolutionary accidents have placed us in that privileged position, so we had better make the most of it and look after this precious planet.

Biodiversity

There are between 5 and 10 million different species of plants and animals inhabiting the Earth, and a very much higher number have existed but are now extinct. (Perhaps some 99 percent of all the species that have ever existed are now extinct.) Such is the incredible biodiversity of the planet. Yet all life-forms had a common origin. Evolution produced organisms that can survive in ice, in bubbling mud pools, on the floor of the deepest ocean, and at the outer reaches of the atmosphere. Water, heat, and time (plenty of it)—they are important basic ingredients in the cookbook of evolution.

In recent years, scientists have become increasingly concerned that the pool of genetic information in animals and plants, the diversity of which is the only guarantee of survival, is foolishly being widely reduced. For example, about 100 acres of the tropical rain forests, the most biologically diverse regions on Earth and the source of many medicines, are being destroyed every minute of every day. At the present rate of destruction, the rain forests will disappear in 25 years, with profound effects on the climate as well as global ecology. The rain forests cleared for farming often expose soil that is poor agricultural land. Scientists suggest that if the present mismanagement of the natural environment continues, some 1 million species will be lost over the next 25 years, the majority never having been tested for possible use to humans. All this is supposedly in the name of human progress!

There is no escaping the evidence that humans are to blame for the current mass extinction of species. Of the twelve mass extinctions of species found in the fossil records of the past 800 million years (the limit of the convincing fossil record), five have destroyed the majority of species existing at the time. Although the mass extinctions of the past may have been caused by extraterrestrial catastrophes, this time humans are the undoubted culprits. Natural habitats have been destroyed. Species have been hunted to extinction, for food and fur. Plants have been destroyed with spade and blade. Land has been laid waste by fire and flood. Indigenous species have lost out to hardy newcomers introduced by humans. The use of pesticides has spread death and disease to unintended victims. Atmospheric and river pollution drives species to cleaner, but more dangerous, environments. Food supplies are depleted by human competition. Even human vanity has played a part, with birds killed for their decorative feathers, and plant and animal oils used for cosmetics.

Legislators introduce laws to protect endangered species and conservation regions are set aside, but the losses and slaughter go on. We have proved to be poor stewards of this fragile planet. We are squandering the natural capital of the planet at such a rate that ecological collapse looks a frighteningly plausible scenario for the future. Forget about horror stories of extraterrestrial catastrophes or nuclear Armageddon. And forget about optimistic hopes of humans eventually colonizing space, and spreading humanoid culture to the backwaters of the cosmos. Terrestrial implosion looks more probable than extraterrestrial expansion. The L factor in the Drake equation again looks frighteningly short, if humans are anything to go by.

Observation

It is worth reflecting on Earth's life pattern. After the rapid emergence of life on the Earth, once the conditions were suitable, 3 billion years of ordinariness descended on the planet. Evolution got stuck in a massive rut (from our point of view; for bacteria, it was very much business as usual). Life seemed to be going nowhere in particular. Things in the oceans were comfortable for a diversity of simple species. Then about 530 million years ago evolutionary boredom was replaced by creative magic. What started the grand opera of Darwinian evolution, during an amazing 5 million years that transformed the planet? Quite simply, we do not know, but once the Cambrian explosion of new increasingly complex life-forms was started, there was no holding it back.

If only we knew what triggered the Cambrian explosion, we would be in a much better position to decide whether "slime" could progress to complex life on a planet of a distant star system, and thence to ETI.

The fact that nothing too significant happened to simple single-cell systems in the Earth's oceans for 3 billion years suggests that the "trigger" starting the race from simplicity to complexity is special, and exceedingly rare. Even when the trigger was pulled, it was the very special characteristics of the Earth described in Chapter 4, plus a sequence of unpredictable events, that allowed humans to emerge as the prime species. There could have been so many outcomes different from even those random events we do understand. What about a sequence of chaotic events on some distant planet, we know not where, shaping the evolutionary destiny of some complex life-form, we know not what? If we have experienced so much difficulty in trying to understand the evolution of life and the emergence of intelligence on Earth, then it is easy to conclude that the path to ETI in the depths of the cosmos will be exceedingly complex and extremely rare.

Humans only just made it by the skin of their teeth. If mammals had not been given the evolutionary fast track by the demise of the dinosaurs, then primates could not have emerged in the leading position in the race for evolutionary supremacy. If the primates had stayed in the trees, then Earth would have remained no more than a spectacular nature reserve.

Judging by the peculiar circumstances that defined the evolutionary path to humans on planet Earth, the chances that intelligent creatures capable of developing advanced technologies will evolve elsewhere in the cosmos may be very small. It is probable that ETI is out there somewhere, but it will be rare indeed and difficult to find. Even if ETI is extremely rare, detection would be of such profound consequence to our understanding of the meaning of life and our place in the cosmos that the search for ETI should be seen as one of the key challenges for science.

Our conclusion that ETI is rare—based on the need for special planetary conditions, as explained in Chapter 4, and the random set of events that shaped our biological destiny, as outlined in this chapter—is exactly in accord with the fact that four decades of SETI have failed to make a positive detection. The conclusions to be drawn from addressing the SETI and McCrea questions turn out to be the same. ETI is out there somewhere, but is sufficiently rare that the SETI surveys may be long and challenging. However, we have the technology and the scientific talent. But do we have the patience and will to see through this challenge to pass on a new understanding of life in the cosmos to our descendants?

PART C

The Fermi Question

•

If they are there, why aren't they here?

6. Spaceships from the Stars

"Truth is stranger than fiction, but it is because fiction is obliged to stick to possibilities; truth isn't."
Mark Twain

Here in Part C we come face-to-face with the scientific "giggle factor." Most serious scientists do not believe that "UFOs" (unidentified flying objects), real or imagined, have anything to do with them. Because science has chosen not to get involved, the topics of UFOs and possible visitations from aliens have been hijacked by the tabloid press and the pseudoscience fringe. This has made these topics look even more "wacky" to scientists, and make them even *less* likely to become involved. Yet viewed from the basis of physics, nothing precludes interstellar travel. Cost might be a problem, culture may be an issue, and time scales are certainly a disincentive, but the laws of physics do not get in the way of robot interstellar craft (although biology may not allow habitation in a craft). Until science is prepared to engage in serious debate about UFOs and claimed visitations from aliens, there can be no prospect of dislodging pseudo-science from the field.

Part A reviewed the commitments and practices that define SETI. We noted its commitment to radio search as a preferred form of experimentation, and the theoretical commitments embodied in the Drake equation. We also noted its lack of success (thus far). In this chapter, we focus on another of its unstated commitments: its implicit rejection of UFOs as candidates for sightings of extraterrestrial spacecraft and beings on

Earth. This is a significant commitment. The majority of non-scientists, conditioned by fiction and the popular press, associate the search for ETI almost entirely with UFOs and not with some specialized subdiscipline of radio astronomy, although perhaps the movie *Contact* has helped to correct that opinion. Yet supporters of SETI by radio search, the program to which professional scientists almost universally gesture when quizzed on the existence of ETI, reject the topic of UFOs as having nothing to do with them. As a rule, the scientists involved in SETI do not merely reject the observation of UFOs as "not my business," in the way they might the prediction of earthquakes or many other respectable scientific subdisciplines; they reject the very possibility of a respectable science that studies UFOs, a scientific "ufology."

The position adopted on UFOs by the SETI Institute is typical of that of the scientific establishment. On their Web site they proclaim, "...you may wish to reflect on the fact that if there were interesting, verifiable evidence that extraterrestrials were visiting our planet, tens of thousands of university scientists would be busy investigating this idea. They're not."

Forty years ago the same could have been said of SETI by radio! However, the SETI pioneers faced down the giggle factor and legitimized their subject. Scientists need to be prepared at least to discuss the UFO phenomenon, if the public is to be better informed about the issues.

Although the study of UFOs is now rejected by nearly all scientists, it was not always so. Following the dramatic growth of claimed UFO sightings after World War II, a significant number of scientists showed interest in these reports. This interest climaxed in the Condon report of 1967–68, based on a vast investigation of countless claimed sightings. The study was made by distinguished physicists, headed by Dr. Ed Condon, and funded by the U.S. Air Force. The report came to a staggeringly damning conclusion about the sightings, not only dismissing every one of them, but coming to the conclusion that further investigation of UFOs would be of no interest to any science except the study of the psychology of mass hysteria. As influential as the report itself was the favorable response it received from the experts of the field. Philip Morrison, the SETI pioneer, reviewed the report in *Scientific American,* writing, "One comes away edified, amused, admiring and well satisfied. ... Science is the stronger for this sincere and expert effort to deal with a public concern."

That was more or less that regarding concerted scientific investigation of UFOs. A few individuals, most notably J. Allen Hynek, did persist in studies, but their lonely efforts failed to produce any evidence powerful enough to get UFOs back on the scientific agenda. The military did keep records of UFO sightings, and sought explanations in terms of well-understood phenomena. The giggle factor came to dominate the response from science to UFOs. The intellectual vacuum left by science was all too quickly filled by the unscrupulous purveyors of fantasy.

This chapter seeks to explore and explain the rejection of a ufology by scientists, particularly those who are involved in SETI. We do not do this because we believe that it is necessary to reevaluate any of the evidence that Condon rejected (and the endless similar material that has been made public since). Rather we believe that exactly the same presuppositions that justify SETI can be used to justify an extended range of searches, possibly including a form of scientific ufology. This is likely to be met with incredulity by many of those sympathetic to SETI, but we ask them to bear with us because the case is worthy of consideration. We must emphasize that we do not accept the numerous sightings of UFOs and aliens, as reported in the tabloid press and by the pseudoscience fringe. All too often these represent the mischievous peddling of fantasies and untruths. Even at their best they certainly come nowhere near minimal standards of scientific objectivity. That should be seen as a challenge for science—not simply to dismiss these efforts, but to do better. Science should provide explanations (in terms of natural phenomena) where they can, and expose fraud and fantasy where they are obvious. Only then will the small number of cases that might, just might, be worthy of a fuller and more open-minded scrutiny be identified. Then we can start setting an agenda for a legitimate scientific ufology, as we will describe in the next chapter.

The first few parts of this chapter are concerned with arguing that any appropriate approach to studying UFOs could be scientific. (The work of J. Allen Hynek and the small band of genuine UFO experts is scientific, but they represent a faint voice of reason among the shouted absurdities of the pseudoscience fringe.)

The defense of a particular domain as part of science is significantly different from simply outlining the scientific consensus on a particular issue, as we were able to do in Parts A and B. It involves engaging with the philosophy of science, a discipline very different from, and in general very much less well organized than, any area of science. It might be said of the philosophy of science, as it has been said of economics, that if you ask two of its practitioners a question, you are likely to end up with at least three answers! Fortunately, by limiting our attention to the case in hand, rather than by trying to draw general lessons, we can simplify things. However, the reader should be aware that in seeking to understand why ufology is not presently regarded as a science by the vast majority of scientists, we are taking a first tentative step into a vast controversy. That controversy centers on what exactly distinguishes a science from a nonscience, and what is special about those disciplines that are accepted as being science.

The proposition that any study of UFOs could be scientific is importantly different from the notion that scientists should study UFOs. The situation where there is no study of UFOs at all would satisfy those who believe the first, but not those who also accept the second. The second half of this chapter, and most of the final chapter, consider the question of whether science should be interested in UFOs. We have an unconventional answer to this question, since we argue that many of the same principles that can be used to justify SETI by radio (which we enthusiastically support) can be used similarly to justify a scientific form of ufology. This might seem a radical position to take, and in a sense it is, but it must be distinguished from a more radical position still that we have no wish to endorse. We do not claim that there is any evidence that there really are alien spacecraft in the vicinity of Earth. Indeed, we think the possibility that any of the reports of UFOs are visiting aliens is extremely unlikely . However, we do believe that visitations (at some distant time in the past or at some time in the future) are genuinely possible, and a significant enough possibility to be worth serious scientific consideration. The exact form of the investigation proposed will emerge during the course of our discussion, and an extensive description of it is the subject of the next chapter.

We lend our support to a scientific ufology with some trepidation, recognizing that it will immediately trigger incredulity in some readers. We would ask for thoughtful consideration of the issues, since they have certainly swayed our opinions (despite the fact that we started out as skeptics)

Science, Objectivity, and the Investigation of UFOs

There is a distinction between the study of UFOs as it is currently practiced, and our proposed future science of ufology. While we cannot specify exactly what is involved in that future science yet, we can say that it must be very different from the vast majority of contemporary studies of UFOs by the pseudoscience fringe. We can see this by considering just one important standard that science sets itself.

The scientific method demands that all claims of observations must be open to test and assessment by other practitioners of the science in a particularly rigorous way. In many cases, this can be accomplished simply by other research groups repeating the observation and experiment. If these other groups persistently fail to generate the results claimed, then the claim is dismissed. It was this standard that led to the rejection of claims of the discovery of "cold fusion" a few years ago. As a rule, the mere threat of this cross-checking is enough to make the initial discoverers sufficiently cautious and careful that few major mistakes are made. Researchers are careful to recheck their results before publishing them. Often the initial result is found to be incorrect, but in a way that can be explained by later researchers. After all, science is a pretty challenging business and anybody can make a simple error in measurement or interpretation during research. Honest mistakes are an inevitable part of scientific progress at the cutting edge of creative endeavor. Science can make progress through understanding the nature of genuine mistakes. More serious errors that cannot be clearly understood as a "mistake anybody could have made" can affect significantly the reputation of the scientist or research group involved. After a few such errors, the further work of the group is likely to be viewed with suspicion or worse, ignored. A suggestion of fraud is more serious still. Any substantiated allegation of a "cooked" (that is, deliberately fabricated) result is liable to end the career of the scientist involved. Scientists who initially accepted the fraudulent result as real will receive some reprobation, even if they were not in any way responsible for the concoction of the result. The essential independent verification of any scientific result has kept science honest, and is at the heart of the SETI protocol. The SETI scientists will not announce a positive detection unless they are absolutely convinced of its authenticity.

This basic standard of independent verification cannot be carried over in a straightforward way to the study of claimed UFO sightings. The majority of claims concern nonrepeatable observations, where someone happens to be on the spot to make certain observations. Yet the same minimal standards of objectivity as required by science should still be observed. Data offered must certainly be very much more than the anecdotal reports that constitute most of the claimed UFO sightings, and there must not be even the slightest hint of deception. Where evidence is offered, it should consist of clear results from well-calibrated instruments, rather than the sort of infamous "blurry photo" that is standard for UFO reports. There would need to be minimum standards for any genuine science of ufology.

The standards required by the scientific method are not merely arbitrary. They are an essential part of science's attempt to establish robust knowledge about things that are not simple to understand. Error is all too easy. That even the very greatest minds are all too prone to error is clear from many episodes in the history of human knowledge. It took until the work of such figures as Galileo and Newton in the seventeenth century to achieve a science of mechanics that had any real power. For many centuries previously, minds of great ingenuity had not been able to improve significantly on Aristotelian physics, a form of organized common sense. For all the sophistication of Aristotelian thought, and those that accepted it, it was deeply mistaken.

A case resembling current sightings of UFOs involves the reports over the centuries of living specimens of some extinct species or other. Some of these are simple fraud (for example, the Piltdown Man skeleton), some are fantasies (sightings of unicorns or mermaids), others are obvious mistakes, while a few may yet turn out to be correct, even if unconfirmed at present. Yet respectable warrant for scientific belief is not forthcoming.

The cross-checking of the scientific community is not the only instance where a social process is used to help ensure that the truth is identified. Consider the process of the law courts. Rather than simply allowing mob opinion to dictate how punishments are handed out, cases must go through an involved procedure of argumentation and judgment, and even then miscarriages of justice are still all too common. Yet the vast majority of the contemporary studies of UFOs fail to meet the minimal checking standards of such judicial investigation, let alone scientific standards. At best, the vast majority of them are of the same standard as poor investigative journalism, based on hearsay, opinion, and gossip. And, as they say, you should not believe everything you read in the papers. A small number of the reports are of a rather higher standard, and we will discuss this material in the next chapter. These more serious analyses are the first early strands of a genuine science of ufology.

Much of the contemporary discussion of UFOs is ignored here, since we are interested in what science has to tell us about ETI, and the contemporary speculation about UFOs by the tabloid press and others is simply not science. It is a sad indictment of science that the shelves of bookshops are filled with books on UFOs written by nonscientists, which are largely based on sensationalism rather than reason. The authors of many of these UFO books are profiting from an eager, but gullible readership that is too rarely given the option to learn about scientific truth.

Of course, there is no reason why we should rely singularly on science as a source of knowledge. Such a position is often criticized as "scientism," and is one of the bugbears of the humanities academic establishment. The scientific mode of thinking is not universally applicable. Consider, for instance, the subject of history. There is no doubt that the vast majority of what is published in history books comes nowhere near the standards of objectivity required in science. Nor does it worry historians when there is a persistent disagreement between them. By contrast, it is a serious concern for scientists when they are unable to reach a consensus on a matter of substance. Could the study of UFOs live by the scholastic standards of history rather than those of science? There is no simple answer. In part, it is a matter of personal sentiment. Some will be willing to believe that UFOs are alien spacecraft simply based on a personal hunch, or the feeble import of urban myth. Others might be willing to accept the reality of UFOs on the basis of vaguely responsible investigative journalism. For our part, we believe that science is the most reliable guide to issues of this kind, and therefore any serious study of UFOs should be scientific. It is very hard to offer any probative argument on this score, but there are good reasons for it, not least among these are the standards of objectivity and error avoidance that the scientific community imposes. The standards of science and the well-established scientific method remain our best bet for determining the truth on UFOs. As scientists it is hardly surprising that we take this line. But history is on our side, as history will show that it was science that was able to demystify the likes of alchemy, sorcery, and astrology. Science must now demystify UFOs. The demystification of alchemy and astrology did lead to legitimate chemistry and astronomy. There will be similar rewards in demystifying UFOs, in improving human understanding.

Our instinct is that science will be more informative on the existence of ETI than will a history style of scholarship. This book is partly intended to prove exactly that point, indicating that it is possible to develop a deep perspective on ETI from the different scientific viewpoints adopted in Parts A and B, and here in Part C.

The application of scientific theories developed for completely different purposes has turned out to be very informative about ETI. It may not yet allow us to answer for certain the key question of whether ETI definitely exists, but it does allow us to argue with conviction that ETI is highly likely to exist. Moreover, it allows us to undertake reasoned arguments about how common ETI might be and where it might be found. SETI depends on this.

No inquiry can take place completely in a vacuum, without any theory against which to test the data. As we will see in detail in the next chapter, a scientific ufology is provided with much of the requisite framework by well-confirmed scientific theories. A historical inquiry, on the other hand, lacks any such obvious framework. It can be modeled along the lines of a martial history, a biographical history, a social history, an ethnic history, or any of the other standard ways of structuring a historical investigation. It is no coincidence that so much of the contemporary pseudoscience investigation of UFOs is structured along the lines of a conspiracy theory. This is because a conspiracy theory is one of the very few available ways of trying to structure the available evidence into some sort of coherent narrative. Real science just cannot cope with the hearsay, speculation, and fantasy at the core of the pseudoscience coverage of UFOs, so conspiracy theory provides an appropriate smoke screen for ignorance. Such a framework is also the natural tendency of another alternative nonscientific approach, a style of investigation modeled on the techniques of a criminal investigation bureau. In the absence of a fully scientific approach, the approach shown in the popular television series *The X-Files* has been accepted by many as an alternative. However, such an approach centered on a conspiracy theory suffers from the severe fault that its central presupposition (that there has been a government-inspired conspiracy surrounding UFOs) is very likely to be false. Indeed, even if some past UFO sighting had been an alien spacecraft, why should there have been a governmental conspiracy of any form associated with it? Given the generally chaotic and disorganized state of many governmental organizations, we find the idea of any conspiracies beyond the very specialized and local extremely unlikely. But discussions of claimed conspiracies and cover-up, from Kennedy to Roswell, from Diana to the Bermuda Triangle, all lie outside the scope of this book. Enough has been written about conspiracy theories. Conspiracy theory can be entertaining if presented as fiction, but it is a dishonest distraction if presented as "fact."

One further weakness of any approach that does not rely heavily on scientific theories is that they may well be misled by the all too many UFO sightings that turn out to be entirely identifiable as natural phenomena. Many a UFO sighting, even one described with apparent sincerity as closely resembling something like a Hollywood-style space battleship, has been subsequently associated with an aurora, a meteorite, a weather balloon, a rocket launch, a distant lighting storm, or one of a host of other perfectly well understood natural or human-made phenomena. In past times, religious visions, described in considerable detail, have been shown to have explanations that prove to be plausible in terms of natural phenomena. Many UFO sightings may be the twentieth-century equivalent of medieval religious visions, or Victorian sightings of elves and fairies. Whereas, in general, scientists will be able to identify cases that turn out to be sightings of fully understood natural phenomena, these may well be confusing for alternative approaches unaware of the range of legitimate scientific explanations available. Of course, it would be possible for a more forensic or historical approach to rely on a scientific filter as to which cases to consider. We hope it is clear that whatever else the contemporary investigation of UFOs is, it is certainly not genuinely scientific in its popularly presented form. Those who favor the contemporary investigation would be better off with the scientific study of UFOs we are arguing for. We do not pretend that it is decisive or that science will necessarily convince everyone. But our belief that science is the only acceptable approach to ufology explains our decision to ignore the vast repository of UFO folklore and conspiracy theory that has accumulated over the years. It also supports the contention of almost the entirety of the scientific community that the contemporary investigation of UFOs does not qualify as scientific (with the exception of one or two studies, discussed later). However, our rejection of the contemporary investigation is not intended to simply leave a vacuum; we believe that scientists should favor a genuine science of ufology as part of an expanded notion of what constitutes SETI.

Things That Go Bump in the Night

Why is it that most SETI scientists reject a ufology? It cannot be because of the failures they find in the contemporary pseudo-science investigation of UFOs. Professional astronomers find no difficulty totally rejecting astrology while pursuing a legitimate astronomical study of the heavens following the scientific method. However, the majority of scientists appear unwilling to accept a ufology, even if it could meet established scientific standards. What is the basis for their rejection?

One standard reason for rejection—that the study is a "waste of time and money," an argument scientists often offer about a certain kind of "big science" project, such as the manned space program—does not seem to apply.

Another standard reason scientists give for rejecting a given domain as a possible subject for legitimate scientific inquiry is that the activity supposes physically impossible operations. This is the criticism that scientists offer of astrology (if taken as more than an amusing game of guesswork). The idea that the positions of the planets can influence human affairs in any very specific way is straightforwardly impossible, given the well-confirmed cannons of physics. The mechanical presuppositions on which physics bases its own very effective descriptions of the motions of the planets and their satellites exclude any such influence. It is not that modern scientists have spent time and effort assessing whether events correlate in any regular way with planetary motions. It is that the sort of terms that astrologers use to describe their predictions are just not those that the scientific community will accept as part of a scientific theory. There is no place for being "lucky in love" in the mechanics that rules the motions of the Solar System. Actually, in the case of astrology this rejection is strengthened by the way that the very different approaches of astrology and mechanics were entertained by the predecessors of modern scientists for many centuries. Given that the inspiration for the systematic studies of the heavens was probably that the patterns observed in the stars allowed one to predict the coming of the seasons, it is not surprising that the human mind inferred that other things might also be predicted by the changing patterns in the heavens. Hence, astrology had some prescientific credentials, until the sixteenth century when the mechanics-based explanation of the motion of the planets gained acceptance. Both historically and conceptually, mechanics and astrology are rivals, and mechanics is one of the most successful of the sciences, whereas astrology is no more than an annoying (although sometimes amusing) distraction.

A similar point can be made with respect to the controversial case of parapsychology. Many scientists, particularly physicists, are unhappy when this discipline claims the same legitimacy as other sciences. In part this is because parapsychology is associated with some very dubious traditions of nineteenth-century science, which investigated seances, fairies at the end of the garden, and the like. Yet in its contemporary manifestation, parapsychology has a more acceptable approach, often having a very skeptical and deflationary attitude to supposed psychic phenomena. For instance, in the investigation of the phenomena of extrasensory perception (ESP), parapsychology has a well-established set of methods, closely related to those of conventional psychology. A psychologist might rely on the use of flash cards and statistical correlations to test the limitations on memory and pattern discrimination of a single individual. Similarly, the parapsychologist makes use of two people, unable to communicate by conventional means, and tests the statistical correlations between their responses to see if any psychic interaction is taking place.

The physicists' problem with this sort of approach is not that such experiments fail to be sufficiently rigorous, but that the physicists know that such correlations could never occur. Physicists understand the sort of interactions that take place in nature to transfer information, and they believe that no such interactions are available to the subjects of the parapsychology experiments. If humans produced natural radio emissions, or if the supposed link was via airborne pheromones or other chemical messengers, then it would be a completely different matter. However, humans do not have built-in radio transmitters, nor could pheromones carry the level of information that is involved in the tests. Thus, it seems the physicists are being asked to consider a completely new kind of interaction in nature, aside from those discovered through the investigations of energy and matter. They regard this as completely unacceptable, since all that is known about physics, which can describe the world with such clarity, does not require some new form of interaction. An additional interaction to account for ESP is no more acceptable to the physicists' way of doing things than would be the existence of ghosts. We do not need to try to answer this issue, however, for whatever else UFOs and aliens are like, they are not like ESP. There is no violation of the well-established discoveries of physics involved in the supposition that ETI exists and has a technological capability that would allow interstellar travel, if not of the ETI itself, then at least of ETI's robots. Life elsewhere in the cosmos would be much the same as life on Earth as far as the principles of all sciences are concerned. The basic laws of physics and chemistry (and therefore of biology) are universal. If life elsewhere is to become intelligent enough to get involved in the business of building spaceships, then some fairly improbable evolutionary events will have had to take place, but nothing that stretches credibility. Humans show how far the physics, chemistry, and biology can be taken, and an extension of this to ETI with a technology allowing interstellar travel presents no problems as far as the fundamental theories of science are concerned.

There can be no question that there is a deep skepticism among many scientists over the possibility of interstellar travel. Frank Drake has been particularly vehement in his criticisms of supposed sightings of UFOs, as was Carl Sagan. There is also a similar rejection of interstellar travel in Condon's summary report of the study he chaired. We can understand their cynicism in terms of the explanations for UFOs proffered by the pseudoscience fringe and the tabloid press. However, we believe their cynicism is misplaced in terms of relating the fundamentals of science to the possibility that ETI will have the capability to travel between the stars. Humans have already built robot spacecraft that have now left the Solar System, after fly-by missions of the outer planets.

Can ETI Build a Spaceship?

We have already emphasized the extreme distances between the stars in the context of how hard it is to send a radio signal between them. It would be much harder to send a space vehicle across those vast distances between the stars, and much more expensive. The richest nation on Earth spent a vast fortune putting a manned spaceship on the Moon, which is only a tiny fraction of the distance to even the nearest star. Very much greater journeys would be needed to traverse the Milky Way Galaxy, which is necessary if, as we concluded in Parts A and B, intelligent civilizations are rare and therefore separated by vast distances.

In considering interstellar travel, it is necessary to assume that speeds very much faster than those presently achievable for human-made spacecraft might be possible. At the speeds at which our interplanetary probes travel, it would take over 50,000 years to get to the nearest star.

Proposed interstellar spaceships fall into two categories: faster than light (as favored by science-fiction writers) and slower than light (as dictated by the laws of physics). The former category has the significant problem that it is excluded by most readings of the laws of physics (at least as currently understood), and the latter has the disadvantage of taking a great deal of time to get to a destination. Faster-than-light spaceships are invoked by most science-fiction movies, as "warp drives" or "hyperdrives." It is simply not acceptable in fiction to have the characters taking tens of thousands of years to travel to their destination. However, faster-than-light travel is forbidden as a consequence of the theory of *special relativity*.

Special relativity is one of Einstein's great contributions to physics and is an essential component in the unification of mechanics with electromagnetism. It is a theory of extraordinary power based on just two main assumptions: the light postulate, that the speed of light is invariant with the speed of its source, and the famous relativity principle, that the laws of physics are the same in all inertial frames of reference. From these assumptions the theory derives the so-called Lorentz transformations, which are descriptions of a space-time that is distinctly different from the Euclidean space-time that is familiar from earlier theories. The new, so-called Minkowskian space-time described forbids one from doing all sorts of things that one would assume from everyday experience. In particular, it involves a very complicated notion of simultaneity. Strange effects occur in relativity. Clocks in motion are slower than those that do not move, and run more slowly the faster they are moving (although one must always add in relativity, relative to the motion of the observer). The theory also has some unprecedented physical implications, in particular the equivalence of mass and energy, famously expressed in that best known of all equations in science $E = mc^2$ (which is even more famous than the Drake equation, although considerably more difficult to derive). The equivalence of mass and energy is most dramatically demonstrated by nuclear weapons, which convert small amounts of mass into unimaginable amounts of energy with horrific violence. The theory is also of interest to the designers of spaceships, because it shows that the tremendous amounts of energy required to accelerate spaceships to extremely high velocities are available in the form of mass. Safely harnessing that energy is a problem, as the human experience with nuclear power shows. However, if mass energy equivalence helps the spaceship designer, then special relativity also poses the biggest challenge. It imposes a cosmic speed limit. Nothing is to travel faster than the speed of light. What is more, the energy input required to accelerate the spaceship increases sharply as one nears the speed of light. Even though the speed of light is almost unimaginable compared to the sort of speeds we are familiar with, the cosmos is so large that the time scales involved in crossing the Galaxy, even if one could reach a high fraction of the speed of light, are measured in many tens of thousands of years. Despite the extraordinary predictions it makes, special relativity is one of the best-confirmed theories in all of science.

If special relativity, one of the best-established theories in all of physics, forbids faster-than-light travel, then why did we even bother to mention the possibility of a faster-than-light spaceship?

The fact is that special relativity is not the only theory in physics, and its relationships to the others are sufficiently complicated that there might be some faint hope of ways around its dictates. For a start, there is the relation between special relativity and general relativity. General relativity is the theory Einstein developed as a generalization of special relativity. It describes a space-time that departs still further from the prerelativistic. The bizarre Minkowskian space-time of special relativity is now curved, and the curvature allows Einstein's theory to provide a successor not merely to Newton's theory of mechanics, but also to his theory of gravity. The presence of matter (or energy) affects the structure of space-time such that the motion of nearby matter is changed and its path follows the curvature of space-time. It has been suggested that this curving of space-time might enable the possibility of effectively faster-than-light travel between two points without ever actually violating the speed limit imposed by special relativity. It needs to be stressed, however, that this suggestion remains highly speculative, and essentially still science fiction despite serious theoretical consideration. Faster-than-light speed might be accomplished in two ways. The first is that general relativity allows the existence of a range of bizarre objects that occur as natural phenomena, such as black holes. The conventional theoretical descriptions of such phenomena allow for unlikely happenings such as time travel, which would enable spaceships to take, as it were, short cuts through space-time. This interpretation of general relativity has been challenged by some theorists, and the question remains unresolved. The second way is the possibility of distorting space-time by the presence of matter or energy; this has suggested to some that it might be possible to systematically distort space-time so as to carry with it a spaceship. Thus, the spaceship would remain at less than the speed of light with respect to its local space, but that very local space would be moved relative to the rest of space-time. This is not at all easy to visualize, but a loose analogy is water current. Relative to the water, we can only row at a certain speed, but if we are rowing downstream, the current will generate a faster speed relative to the bank. Imagining how to create a favorable "current" in space-time is by no means straightforward. Indeed, it is probably fair to say that no theorist has demonstrated that it is even mathematically possible, let alone physically possible. Faster-than-light travel thus remains in the realms of science fiction. But one day, science may change all that. Stranger things have happened. The equivalence of mass and energy would have appeared to be science fiction to nineteenth-century physicists, but at Hiroshima it became a crucial reality to twentieth-century peoples.

In addition to the complexities due to general relativity, there are also some associated with the relationship between special relativity and quantum mechanics, the strange theory of how light and matter interact on the scale of atoms. Quantum mechanics does not combine well with relativity, and it is still a major challenge for theorists to complete a full reconciliation. It is possible that this reconciliation, which if it includes general relativity is often described as a theory of quantum gravity, will enable the local speed-of-light restriction to be violated under certain very special conditions. Since we do not yet have the reconciling theory, it is impossible to say whether such conditions might exist, what they might be, and whether there will be any chance of exploiting these possibilities practically. For instance, it may be that such conditions only occurred in the very earliest stages of the development of the universe when energy densities were unimaginably high. If this is the case, then there is no chance of them ever being used by ETI, since no life could exist under such conditions. The whole field remains uncertain. However improbable it might seem in terms of current scientific understanding, faster-than-light travel cannot be dismissed entirely as mere fantasy or impossible in terms of the physics. The theoreticians may yet have some surprises in store for us.

In discussing the motivations and expenses of interstellar travel, we will consider only the case of slower-than-light travel. Faster-than-light travel seems improbable, but slower-than-light interstellar travel is certainly plausible in terms of the laws of physics. Nevertheless, it is important to emphasize that even in the case of slower-than-light interstellar travel, the velocities involved will need to be considerably beyond anything achieved so far by humankind in their exploration of the Solar System.

Unlike faster-than-light travel, we do have a reasonable idea of how spaceships traveling at speeds of, say, a tenth of the speed of light might work. There have even been several attempts to outline designs of interstellar vessels, such as the British Interplanetary Society's Project Daedalus in the mid-1970s. The intention of the project was to design an uninhabited probe that could reach Barnard's star, some 6 light-years away, inside 50 years. The result of the design was a nuclear-powered behemoth of 50,000 tons, almost twenty times the weight of the Saturn V rockets used to power the Apollo Moon missions. Nor was the choice of nuclear power merely the gratuitous fashion of the age. There was no chance of using chemical fuels to power a spaceship to the high fractions of the speed of light implied by the 50-year limit, because in a chemical reaction only a very tiny proportion of the mass of the fuel involved is released as free energy. In a nuclear reaction, a much higher proportion of the mass is released. Scientists really looking for results should choose matter-antimatter collisions in which all the mass energy of the reactants is released. The problem is that the more energetic and unstable the fuel, the harder it is to produce and use safely. Half a century of experience with nuclear power has not been totally reassuring on that score.

The huge energy potential of nuclear reactions makes travel at a substantial proportion of the speed of light at least conceivable. However, the energy requirements for interstellar travel are quite mind-boggling. Frank Drake pointed out that the energy required to send a group of 100 colonists to another star is equivalent to the total energy needs of the United States over a typical human lifetime. It is hard to contemplate such an enormous energy demand; however, we can envisage it being met by highly efficient nuclear fusion power sources. The energy economics might not make sense to us, but we would be foolish to try to imagine the energy economics of a civilization considerably more advanced than ourselves, as ETI is likely to be.

Of course, the engine is only the first part of the design challenge. Many other technical advances would need to be made. For instance, at speeds of the order of a tenth of the speed of light, a collision with the smallest of dust particles becomes a potentially catastrophic explosion. The spaceship either needs to be coated in some fearsome armor, or else needs to incorporate some sort of clever trickery to "zap" the incoming projectiles. At these speeds, changing direction too often is not really an option, other than by gravity assist from the stars along the desired trajectory. Slowing down at the stellar destination will take as much energy as accelerating, and again this would probably have to be done with nuclear rockets.

The really challenging decisions would concern the payload. Given that the travel times involved are so long, it would not be possible for humans to complete in a normal lifetime a round-trip to even the nearest star, let alone across the Galaxy. Hence, the notions of cryogenic freezing of the crew, and the like, have been popularized by science-fiction authors. Of course, an ETI might have a much longer life span than humans, but it still seems unlikely, although not impossible, that the ETI life span is anywhere near the millions of years needed to transit the Galaxy at speeds on the order of a tenth of that of light. (Then again, ETI might have a spectacularly shorter life span than humans; time scales are all relative. A "three score years and ten" life span looks sort of normal to us, but maybe not to ETI.) Hence, it seems one can choose between a payload of robots or "frozen" astronauts (if suspended animation ever proved to be possible), or huge colony ships where generations will live and die in the darkness of outer space. However, before such decisions can be made, one needs to have decided exactly what purpose the spaceship is to have. We will discuss this in more detail later. The important point to record here is that slower-than-light-speed interstellar spaceships are possible, at least in terms of the laws of physics. There might be all sorts of reasons why interstellar travel involving ETI, or robots, or (in the very distant future) humans will not take place, but they are unlikely to be because the physics ultimately gets in the way.

An interstellar spaceship is not likely to be built by humans for generations to come. Futurology is an inexact science, to say the least, and such predictions often have a habit of going astray. (A spectacular example of an inadequate prediction is that made by a prominent American newspaper in early December 1903, that man would not fly for a thousand years, which turned out to be wrong by 1,000 years minus 4 days!). Probably the most significant challenge is to get the enormous spaceship required into orbit around the Earth, before it sets off on its cosmic quest. Since any vehicle would almost certainly need to be assembled in space, every ounce would need to be carried into orbit. Perhaps it could be assembled on the Moon, which with its lower gravitational field would make a launch easier, but materials (other than those that could be mined locally) would need to be ferried to the Moon. The international space station currently in its early stages of assembly is going to cost many tens of billions of dollars to put a few hundred tons into space. Any interstellar spaceship is likely to be very much bigger than that, and with a substantial proportion of its mass made up of a highly dangerous nuclear fusion reactor. It will be a long time before any object made by humans reaches for the stars in a serious way. (We are talking here about a targeted stellar probe with the ability to orbit or land on a planet in a distant star system.)

What about interstellar travel for ETI? That is a different matter. For ETI, time is on its side.

Time Is on ETI's Side (or Is It?)

The Milky Way Galaxy is certainly very large, but we cannot really understand the chances of ETI building a spaceship unless we appreciate that any ETI with the desire and capability to travel to other worlds is likely to be very old. Remember that the Milky Way is thought to be about 10 billion years old. Our own Earth is approximately 4.6 billion years old, and has been a living planet for roughly 4 billion years. Compared to anything we are familiar with, these are enormous time scales. A century is one-ten-millionth of a billion years, about the same fraction as 3 seconds is of a year. Yet a century is a long time relative to technological and scientific advancement. The twentieth century has seen the invention of the computer, the airplane, plastics, penicillin, and endless other technical devices and processes that have changed society beyond recognition.

There is no reason why ETI should have become intelligent at the same time as us. The majority of stars in the Milky Way are billions of years older than the Sun; therefore, any ETI could be millions or billions of years older than us (if it has survived), even after taking into account the fact that the early generation of stars in the Milky Way would not have had the right mix of elements for life-bearing planets. It could be that ETI evolved and died a billion years ago. For simplicity, however, let us consider a more conservative estimate, an ETI that emerged a hundred million years ago (of course, there is equally no good reason why a future ETI should not be getting on the evolutionary escalator some hundred million years behind us). A million centuries is plenty of time to make unimaginable advances in technology, especially when we think what our last century has produced. Indeed, for ETI a million centuries may be long enough to achieve space travel.

Although we do not know how to build a stellar spaceship at the present time, with the rate of technological advance it is not too hard to imagine such technology being developed within a further millennium (if the human species survives). A period of a few millennia is an extremely short interval in terms of the evolutionary time scales of the stars. And if one has a spaceship, even one that can only travel at about a tenth of the speed of light, then a million years is enough time to have traversed the Galaxy. Of course, there remains the problem of longevity referred to earlier; however, a commitment to colony spaceships with ETI breeding endless generations gets around that issue, as do robot craft.

If ETI civilizations do exist, then the possibility of technologically advanced and widely traveled ETI would suggest that, far from being something stuck on a faraway planet sending radio signals out into space, as SETI by radio assumes, ETI could already have been here! As noted in the Prologue, the first phrasing of this argument is normally attributed to Enrico Fermi, not entirely coincidentally one of the pioneers of unleashing the power of nuclear energy. He was fond of asking his colleagues the question, "If aliens exist, why aren't they here?" Strange though it may seem, once one has accepted the notion that slower-than-light interstellar travel is possible, and has properly appreciated the relationship between the extraordinary length of evolutionary time scales relative to technological ones, one realizes there is no obvious answer to this question, although some have rushed to the answer that we must be alone in the cosmos. Fermi fully understood the implications of his question, but did not advocate an "all alone" answer. Others were to adopt the Fermi question, however, as the basis of advocating the "all alone" hypothesis.

For convenience, let us use a new term to describe ETI civilizations that have developed an interstellar space traveling capability. We will call them "ETI capable of *interstellar mobility*," or IMETI for short (pronounced eye-meti, yet again providing that rhyme for "yeti"). Our apologies for introducing this new term, which we admit is far from pretty, but we think it is important to distinguish between the ETI advanced enough to transmit radio signals (to be searched for on Earth by SETI), and that group of ETI who have developed a wanderlust. By this time the reader will be familiar with the yeti, seti, eti rhyme, and adding eye-meti (IMETI), also rhyming with yeti, to the lexicon will hopefully not be too confusing.

One of the original intentions of those offering the Fermi question argument was to suggest that since we have not seen IMETI, ETI of any sort must be so rare that SETI by radio is a misguided enterprise. They sought to create the so-called Fermi-Hart paradox between the existence of abundant ETI and the nonobservation of interstellar visitors. (Astronomer Michael Hart took up the Fermi question, and developed it into an attack on SETI. Drake argued with great conviction that the Hart approach was flawed.) The Fermi-Hart argument needs serious consideration, up to a point. We believe it probably does rule out the possibility of the high abundance of ETI civilizations that some of the founders of SETI believed in. However, this conclusion can be equally reached from other directions, as we did in Parts A and B in addressing the SETI and McCrea questions. But we certainly do not support the Hart line that the absence of firm evidence for IMETI means that ETI does not exist—and therefore SETI is fruitless. IMETI could have come and gone, or still be on its way, or just not be interested in "little ol' planet Earth."

We have already discussed in some detail the possibility that ETI might be rare per se in Parts A and B, and concluded that the number of ETI civilizations must be well below the high levels argued for by the SETI optimists. However, neither the arguments from astrobiology nor the Fermi-Hart paradox show that it is impossible for significant numbers of ETI civilizations to exist. By "significant numbers" we mean of the order of perhaps a few thousands in the Milky Way Galaxy, rather than the hundreds of thousands, or even millions, sometimes proposed. Nor, it should be pointed out, have scientists searched overly hard for evidence of the extraterrestrial spaceships whose nonobservation is the first part of the paradox. IMETI may not have made itself too apparent to a casual observer. But if we do not search, we will not find—even if the search does involve a fair degree of hope and speculation about a possible tiny needle in an unknown haystack.

If we reject the necessity of the inference that there must be very few ETI civilizations, then there are four possible answers to the Fermi question. The first is to claim that there must be some sort of technological barrier to interstellar travel. The second is that there may be sociological barriers to development, such as limits on cost, or merely the failure to find an adequate motivation for the efforts involved. The third is to claim that IMETI is rare because long-lived ETI civilizations are rare. This rarity of long-lived civilization must be because they tend to die out young through the perils of nuclear war or the like, as we speculated in Chapter 3. The fourth possible answer is to concede that it is just as possible that there are significant numbers of IMETI as it is that there are significant numbers of radio-transmitting ETI, and we just need to keep searching for IMETI as an extension of conventional SETI. From this, it would seem to follow that a scientifically grounded search for interstellar spaceships of ETI origin is as well grounded as SETI by radio. In other words, if one accepts a scientific basis for SETI, which we most certainly do, then it is surely as equally valid to accept the existence of a scientific basis for a ufology. Despite the giggle factor and the wacky approach to UFOs by the tabloid press and the pseudoscience fringe, if we accept the scientific case for SETI, it would be wrong not to try to establish a comparably sound scientific case for a ufology. That is our belief, although we realize it will not go unchallenged by the scientific establishment.

Very little can be said at the present about the first response, of a possible technological barrier. Certainly, it is one thing to claim that nuclear energetics makes interstellar travel vaguely possible, and quite another to show that all the possible technical problems can be solved (at a cost). Perhaps the problem with particle collisions cannot be resolved, or the effect of all the various forms of radiation to be found in space (not least that generated by nuclear rockets) cannot be shielded within acceptable weight limits. The simple answer is that we cannot know whether it is true or not. Any attitude that requires the existence or nonexistence of technological barriers to interstellar travel must be open to question; a certain careful agnosticism needs to be cultivated.

Motivations for Interstellar Travel

The second of our possible answers to the Fermi question is to claim that while interstellar travel might be technologically feasible, there might be social barriers to prevent it. This is by no means an outlandish position; any society that wanted to launch interstellar spaceships would face substantial challenges. The most obvious one is cost. Often in the past, fictional depictions of alien societies tended to give away the author's opinion about where human societies are heading, many being of a distinctly fascistic character. These days, it seems a much better bet that advanced technology will thrive in a free society. However, we cannot know a lot about extraterrestrial economics. An ETI society might function very differently from any of the societies on Earth. For instance, it might not use money, or anything like it. However, there will have to be some means of deciding how the burdens of labor and energy expenditure are distributed. By any reckoning, spaceships are going to qualify as a major expenditure of both. The question of cost is closely bound up with the question of purpose. A culture might be willing to expend tremendous efforts when it is clear that there is some obvious payback, but not on a mere whim. While it is impossible to make any definite statements about how very alien minds might feel about the economics of interstellar travel, we can at least outline some possibilities based on analogies in our own experience. Whatever else a society that might build interstellar spaceships might be, it will certainly have an advanced scientific culture. Otherwise, it would not have obtained the tremendous scientific and technical knowledge needed to build a spaceship.

Every mature government recognizes the need to spend something on science. However, each tends to have conspicuous spending priorities. Science related to military purposes tends to receive a particular priority, and so does that which feeds into health care. Science that promises technologies that could be commercially exploitable can expect substantial backing. However, not all science funding is made with such a clear payback in mind. For some pure science there is no clear payback in terms of increased military or commercial power, or indeed any obvious quality-of-life payback. Consider, for instance, the investigation of cosmology; nobody obviously benefits except the scientists (who are always calling for more money). Even so, there is still a generally agreed sentiment that these things are worth contributing tax dollars to, even if the more extravagant projects cannot always be bankrolled. Where governments do not offer the requisite handouts, private individuals who do contribute are generally held to be doing something worthy, rather than just providing a handout to an already privileged section of society. In part, the motivation for these projects lies in their hidden benefits. One never knows what outcomes pure science might have. It was apparently rootless investigations into the properties of strangely conducting materials that led eventually to the development of transistors. Endless stories can be told of research that was intended simply to answer questions posed by a curious scientist and ultimately turned out to be of tremendous benefit to humanity. But this is not necessarily true of all sciences. It is hard to see how cosmology might have unexpected spin-offs; it either delivers on its own terms (to increase human understanding) or not at all. However, even sciences like cosmology do have some form of payoff. They can be an important part of an active scientific culture and profession. We may not need the results of the cosmologists' work. Nevertheless, we do need the young physicists they train as part of their professional life, a life that they could not lead if they were to function purely as teachers. Equally, people normally can only become professional scientists through the experience of research. This requirement is recognized in the near-universal necessity of a doctoral degree for entry into the community of career professional scientists. We may not need the direct products of the research, but we do need the expertise that scientists develop in problem solving. The idea of a scientific community that makes good by simply imparting the knowledge it has already learned to future generations is not a very likely one. Finally, even beyond these indirect benefits, there is a feeling that these projects enhance our shared humanity, in much the same way that the production of great works of art or grand buildings do. In our cynical age, it is not fashionable to talk about purposes of humanity beyond simple notions of survival or the utilitarian ideal of the greatest possible happiness to the greatest possible number of people. In a bygone age, however, the idea of human purposes, of something beyond the transitory and hedonistic, had a resonance. It is true this is a notion that needs to be kept carefully in check. It has led to monstrous imperialism and totalitarianism; too many lives have been sacrificed for the benefit of our "shared humanity." Yet, once it is put in its place there is still some scope for human purposes. The growth of our ability to understand the nature of which we are a part is surely as fine a purpose as any. We certainly believe it is adequate justification for the support of SETI.

So much for the motivations for investing in science. How might similar motivations be carried over for the matter of building spaceships? It is possible to imagine situations where interstellar travel is of a military or health benefit. Perhaps, the arrival of a robot interstellar probe from one ETI might lead another to feel the necessity of developing an interstellar travel capability. This might conceivably play a role in a snowballing accumulation of ETIs with interstellar mobility capabilities once one had achieved it.

One of the better-known proposed motivations is that interstellar colonization will be required owing to the exponential increase in the population, or even the impending death of the home star. The arguments for both situations, however, are quite weak. Any civilization that has the capability to build spaceships presumably will first have developed techniques for building cities of great size and techniques for growing food within the available space. Even if the surface of the home planet did become too overcrowded or unsuitable because of pollution or the like, the civilization would almost certainly be able to build vast space structures and colonies on the other planets of their solar system, or a Dyson-sphere form of space colonies. These colonies would permit population growth at far less expense of effort and energy than attempts to colonize distant planetary systems. In addition, such proposals seem to underestimate the difficulties associated with colonization, even once one has reached another solar system. Any biological species is naturally tied to its own biosphere by molecular biology. We can digest the sugars that are produced by our plants, and we have resistances to our microbes and the like. Any attempt to colonize a distant planet will require the creation of a "planet like home," either by massive conversion of an existing living planet or the "terraforming" of a sterile one. (We will discuss the process of terraforming in the next chapter.)

As for the death of stars, this is likely to be a problem only in the extremely long term. We have already argued that ETI is most likely to be found around stars very similar to our Sun. These stars have a very long lifetime, 10 billion years or so. Hence, even the very oldest Sun-like stars in our Galaxy are only now beginning to run low on their nuclear fuel. The majority of Sun-like stars still have billions of years ahead of them. In fairness, it must be pointed out that there is not only the danger of the star dying. As we discussed in Part B, even a supernova event some tens of light-years away might well be sufficient to cause the destruction of a planet's ecosystem. However, stars in the regions where supernovae are most frequent are unlikely to have developed ETI populations in the first place, exactly for these reasons. Stars can move in and out of supernova-rich regions, but only on very long time scales. Considering the dangers likely to be inherent in interstellar travel, it would take quite a danger to suggest the development of spaceships on health and safety grounds alone.

Might spaceships be justified as a commercial venture? Science fiction is fond of the notion of the "mining colony," a commercial enterprise based in some nearby barren solar system that happens to be rich in some depleted resource. It seems unlikely that this fantasy can have much basis; an advanced society might well have a demand for an expendable resource, supplies of which could be found in another solar system. However, it seems unlikely that mining another system is the optimum way to solve resource problems. The transportation costs associated with returning mined mass to the home star system means that any product would have to be worth very, very much more than its own weight in gold. It seems likely that synthesis of an alternative to the material in question would be a much better investment; however, mining on other planets within the same star system might be a plausible option, and indeed a necessary step in the construction of a Dyson sphere.

The justifications for the construction of spaceships in terms of ancillary benefits appear to be fairly slim. How about a justification in terms of either pure scientific interest or more general purposes? There is much to be said scientifically for actually going and visiting other solar systems. Plenty of observations of the universe can be made from within one's own solar system. The Hubble space telescope, and other members of the successive generations of space-based observatories, have taught us that much. Future astronomy may well take place using space-based instruments of unparalleled power. However, there is no substitute for close proximity, and there are plenty of things well worth studying that might require a nearby observatory. An obvious example might be the study of the more distinctive characteristics of stars, such as their magnetic fields and stellar activity. We are yet to achieve a clear understanding of these with regard to our own Sun, although satellites have been put into orbit for this purpose. It would certainly be scientifically attractive to have an understanding of the surfaces of stars of various types. A study that required sending nuclear-powered robot probes to observe the characteristics of neighboring stars would be an expensive and long-term research project, but it would produce extremely valuable results. A community of scientists in a society where slower-than-light interstellar travel is possible would certainly be tempted to dream up affordable schemes along these lines. One would imagine these would be robot probes, rather than IMETI visiting "in person."

Of course, the greatest scientific prize of all would be to investigate another life-bearing planet. A life-bearing planet might be spotted using advanced telescopes by virtue of distinctive characteristics of its spectrum. If such a planet were discovered, even at a distance whereby it would take several centuries or even millennia for a probe to travel to it, would it not be tempting to try to send a probe to investigate its biosphere? The scientific results would be gained by later generations than the probe builders. Any ETI, just as we do, would wonder to what extent the features of its own biosphere were paralleled by others. Even if there was no promise of uncovering intelligence on a planet, the investigation of the genetic material of even the simplest organisms descended from a completely different genesis of life would be most significant for understanding the relationship between molecular and evolutionary biology. Unlike the star-studying mission, which could use a probe similar to that which we use in our own Solar System adapted for use in another one, the study of a novel biosphere would require a complex probe with considerable flexibility.

These are mere speculations, but it seems likely that the scientific prizes of the investigation of other suns, other planets, and other biospheres are more likely initial motivations for interstellar travel than any others. However, is there anything to be said for space travel for its own sake? One can conceive of a society where the yearning for interstellar travel has become an essential part of the basic values of the society (as the Moon race was in the United States and Soviet Union in the 1960s). It could be claimed that interstellar travel need not be for anything; it could be good in itself. Such an outlook of human society is portrayed in the popular television science-fiction series "Star Trek: The Next Generation." Of course, our society would have to change a great deal before it came to resemble one that believed it had a mission for good out among the stars. Human values do change; consider the difference between the society and values of Homeric heroes, of the medieval age, of the time of Jane Austen, and of our own day. Such changes take centuries, but as we have already noted, centuries are not something that an ETI civilization is likely to be short of. The two main catalysts to such a change would be the greater significance of space travel in general, something that seems inevitable in the very long run, and the success of the sort of scientific missions we described earlier.

Let us review our distinctly conjectural discussion of possible motivations for making the immense investments required for interstellar travel. It seems most likely to us that any initial traveling will be done with scientific intent, rather than commercial, military, or improving quality of life purposes. If such scientific work, particularly the investigation of alien biospheres, were successful, then it seems possible that over time a civilization will come to prioritize interstellar adventure as a valuable good in itself.

Extinction and All That

The third possible answer to the Fermi question is that ETI civilizations are too short-lived to ever get around to the business of interstellar travel. This position has been taken by several authors, and we included it ourselves in the "medium rare" options in Chapter 3. This position is strongly associated with pessimism about the future of human civilization, driven by the dangers of the nuclear standoff of the cold war or the threat of environmental catastrophe. Some believe that the expected life span of technological civilizations is so low that even radio-broadcasting ETI are out of the question.

We believe that such pessimism could be misplaced as an argument against ETI's advancement in the long term, although it might help in the debate on whether ETI is relatively rare. One might reject the pessimistic hypothesis that advanced civilizations tend to become extinct on the basis of the traditional belief of Marxists and some others that progress and advancement of society are inevitable. We do not agree with this optimistic view. We fear that both nuclear war and environmental damage are all too probable if sensible action is not taken by the world's political leaders. In human terms, these are catastrophic and avoidable tragedies. However, from the long-term standpoint we have adopted, they will be little more than minor blips. We believe that intelligent civilization, terrestrial and otherwise, will tend to be robust. Even catastrophes on an unimaginable scale will not result in the absolute extinction of the species. (The exception to this confidence would be the collision with a giant asteroid. However, in Chapter 4 we argued that the technology now exists to counter such a threat.) The still-substantial community of survivors from global catastrophe would be able, in the long run, to reestablish the species on a more substantial footing. Humans might drive themselves back to the dark ages if they do not take care, but even nuclear Armageddon is likely to leave a few survivors who could reestablish over subsequent generations a new technical order. A substantial body of literature, cinema, and computer games is concerned with such post-apocalypse scenarios. There can be little doubt that life under such conditions would be extremely difficult. However, it takes an extreme pessimist to believe that even if but 1 percent of the world's population survived some catastrophe, the 60 million who remained could not get things going again over many generations. In Drake equation terms, L might be short, but following an extended period of technical oblivion a new technically sustainable civilization might arise with a more extended L value the second (or subsequent) time around. Modern civilizations on Earth did build from the foundations of ancient civilizations that had self-destructed.

This emphasizes the difference between the time scales of astronomy and those we are used to. From the astronomical perspective, unimaginable disasters would be considered little more than blips on the historical progression of an extremely long-lived species. It is true that many species do become extinct, but they tend be those that evolved to inhabit one particular kind of environmental niche in one particular habitat. Intelligent species are extremely unlikely to make themselves vulnerable to the loss of a single resource or the devastation of a single environment. They are not unique in this respect; the rat and the cockroach are even more robust against environmental catastrophe than are humans (a somewhat humbling thought). It is likely that the historical records of any long-lived civilizations that are out there will record some disasters and attendant "dark ages." However, only a closely spaced sequence of serious disasters is liable to cause the total extinction of a robust species.

Bursts of gamma rays of enormous intensity have been detected from distant galaxies. So strong are these bursts that thier only plausible source is the collision of two black holes. Any ETI existing in a galaxy in which such a giant gamma ray burst occurs could be wiped out. It has been suggested that such gamma ray bursts in our own Galaxy eons ago may have sterilized the Galaxy; essentially resetting the biological clock every few hundred years. Perhaps this is the answer to the Fermi question: evolution of life has to start afresh before there is time for ETI to colonize the Galaxy.

Images of ETI

It is possible that the reader will have noticed a distinct omission from our discussion so far. We have not yet offered any real opinions about the nature of ETI, beyond the basics that it will have considerable intelligence and manual dexterity, and that it will belong to an advanced technological society. We have not considered what it will look like or what its attitudes to us might be. This is because we have been principally concerned with what we can know before contact, rather than what we might learn after contact. On a slightly more whimsical note than our previous considerations, we will speculate on the possible appearance of ETI.

Perhaps you have not noticed our deficiency in not considering the appearance of ETI so far. After all, there are all too many images of aliens around, and perhaps one of these has been the reader's subconscious image of our uncommitted "ETI." The majority of common images are derived from movies, since aliens have been a popular theme in recent cinema. In fact, these days it seems that every other Hollywood blockbuster has some reference to aliens, or cosmic catastrophes, and plenty of weekly television series on a lower budget have followed in their footsteps. In many cases, there is very little connection between the alien of the movie and the scientific concept of ETI, simply because the film is actually part of a completely different tradition transposed into an extraterrestrial context. One of the standard archetypes is the alien as predator. If lions, tigers, bears, and sharks are no longer terrifying enough, then how about a fearsome alien that bursts from its victims' chests? The vast majority of aliens of the classic 1950s B-movies are also more in the model of cheap horrors than genuine ETI. Similarly, the *Star Wars* series of movies makes no real use of the notions of difference between the various species of aliens and their respective planets of origin. *Star Wars* was really inspired by many traditions of a more terrestrial nature: Japanese samurai movies, westerns, and the traditional mythologies of many cultures. It is much more akin to a fantasy story like Tolkien's *Lord of the Rings*, or Wagner's operas in the *Ring* cycle. The aliens in *Star Wars* are more like dragons, elves, and orcs in their dramatic role than genuine ETI-type aliens.

Beyond the predator and space fantasy types, some fictions are genuinely concerned with the issues surrounding encountering alien creatures, at either their place or ours. The former is very different from the latter. The classic examples of the "we go find them" genre are the *Star Trek* television and movie series, although there are many notable variations on the same theme. In the case of "they come here," we are looking at something more like *The X-Files* or *Independence Day*. Somewhere in between the two is the memorable and thought-provoking movie *Contact*, which actually highlighted SETI by radio, with the heroine going on a journey to communicate with the senders of a received signal.

Despite the great imagination and variety shown across the genre, these fictions are likely to distort systematically our impression of what ETI might be like.

As noted, authors tend to favor faster-than-light travel for narrative reasons. Similarly, our society's idea of what aliens might look like has been systematically distorted by the requirements of motion picture production. Until the relatively recent development of computer-generated characters, it has been necessary to have all the aliens played by human actors. (A few exceptions involve the sort of stop-motion technology used for *King Kong*, or else the sophisticated puppet technology used to create Yoda in *The Empire Strikes Back* or the lovable *ET.*) In the lower-budget productions, many of the aliens were simply human actors with a little bit of glittery makeup and shiny clothes. Up the scale a bit, extra facial makeup allowed for the addition of distinctive facial features such as the ridges that distinguish Klingons in the later *Star Trek* series. Beyond these are full false heads, which need to be slightly bigger than a human head in order to fit over the actor.

The net effect of this menagerie is that our common perception of an alien is something very similar to humans. They are expected to have an upright torso, to which are attached two arms, two legs, and a rather large head. It is pretty unlikely that any alien will resemble anything like this. Human form reflects a sequence of chance happenings and evolutionary pressures that we cannot expect to be common to all intelligent species. Four limbs may be the popular number among large animals on Earth, but there does not seem to be any necessary reason for this; it is simply a result of the fact that they are all members of a lineage that initially had four limbs, and that mutations that produced a viable species with a different number of limbs seem only to have been possible back in the Cambrian era when life was young. In the biosphere to which ETI belongs, it might be that large creatures belong to a six-limbed lineage, or possibly even a lineage still less like our own form.

An upright bipedal gait is more distinctively human. Indeed, there is considerable speculation as to why evolution through natural selection should have allowed humans to "walk tall." Most plausible accounts connect it to the initial origins of modern humans on the African savannah. An upright posture, like humans' comparative lack of hair, may have been necessary to help with cooling on the hot plain, or to look over high grass after descent from the forests. There is an evolutionary cost in trying to support our outsize brains on top of a long and delicate spinal column, an arrangement that has been compared to balancing a pineapple on a broomstick. We all pay the price for this arrangement with our tendency to suffer from back pain. Unless they have a muscle structure very much better than ours, or else come from a lower-gravity planet, the Hollywood aliens with their even bigger heads are going to suffer more. Perhaps this explains why aliens in the movies always seem to be in such a grumpy mood!

Some scientists have tried to argue that the evolutionary advantages that led to human intelligence, and the need for dexterity to generate technology, mean that ETI has at least a vaguely similar physiology to humans. The arguments tend to stretch the limits of scientific speculation. We simply do not know.

In conclusion, in trying to imagine how ETI might look, we must remember that the requirements of evolution differ from those of Hollywood productions. As some biologists joke, the problem with "little green men" is not that they are little, or even that they are green, but that they are men. This is not a token gesture to feminism, but rather an observation that it is not obvious that separate evolutionary processes would lead to intelligent species that are similar in form to humans. Intelligent beings with the dexterity to develop advanced technologies surely will be out there. But forget about humanoid-type aliens. Suspend all preconceptions. ETI will not look like anything we have ever imagined, which renders even more ludicrous the humanoid-like descriptions of alleged alien abductees, and the unimaginative aliens of Hollywood.

· · ·

Observation

If a long-lived ETI is out there, and were motivated to try a bit of interstellar travel, then what is to stop IMETI from coming here? The answer is nothing but fortune. Remember the important commitment of philosophy that justifies SETI by radio: the scientifically possible is worthy of our scientific attention. Using this principle we can argue that the search for IMETI is also a worthy science. It is important to note that this is decidedly different from a straightforward ufology of the type based on urban myth. For a start, it is driven from the opposite direction. Conventional ufology starts with phenomena, UFOs, and looks at the possibility that they are extraterrestrial spacecraft. We, on the other hand, have come up with a theoretical possibility that advanced ETI civilizations with attendant interstellar spacecraft exist, and propose a scientifically based ufology to look for them on a similar basis that SETI searches for ETI. The next chapter argues that the investigation of UFOs is a part of that study; but it is only a part.

7. Hunting IMETI

"You shall seek all day ere you find them; and,
when you have them, they are not worth the
search."

William Shakespeare

Let us review the position we have reached. We are convinced that the existence of IMETI is a real possibility, and that it is important to remember this possibility when working out how we might search for ETI in general. This is still not quite our promised conclusion, that a science of ufology could be shown to be as valid as conventional SETI by radio. All this is necessarily more highly speculative than what is presented in Parts A and B, and we do not apologize for that. But the reader must remain alert to the transition we have made from a widely accepted SETI in Part A and a soundly based consideration of the relevant science in Part B, to our speculations on a possible expansion to SETI to embrace a scientific ufology here in Part C.

The possibility of ETI being on the move does not guarantee that investigating the UFO phenomenon might be informative. There are three obvious barriers. The first is that IMETI may be busy exploring the rest of the Galaxy and not yet have had reason to travel to our Solar System. The second is that even if IMETI is in our vicinity, how can we say that investigating UFOs is likely to be a particularly effective way of looking for them? Perhaps alternative search strategies would be more effective, or perhaps there simply is no way of getting conclusive evidence that IMETI is in our vicinity. The third possibility is that IMETI may already have come and gone, a billion years or more before our time. The primitive Earth, with its "slime," would have appeared pretty uninteresting, and there has been no cause for ETI to return. The rest of the cosmos may have been of greater interest.

These are all real, if somewhat speculative, possibilities and this chapter will address them in turn. Together they serve to remind us that the space travel argument is not simply a way of supporting ufology from the same principles as SETI. Rather it shows that the principles of SETI support a whole panoply of possible investigations, of which a genuinely scientific ufology is a member. We will structure our discussion of these possible investigations according to the way in which they explore the same possibilities that we saw as blocking ufology: that IMETI has decided to stay at home (in the same sense that SETI nondetection might be because they have decided not to communicate), that they are on their way (in the same fashion that SETI assumes ETI's radio transmissions are on their way), or that they have come and gone (in the same fashion that we considered a range of reasons for SETI nondetections). Lastly, we will consider the proposed scientific ufology itself.

IMETI at Home

Just because they have a technology to travel to other star systems does not mean that any IMETIs are not planet based. Nor does it mean that they will have abandoned radio technology completely, although they might regard it as rather quaint. Hence, there is nothing in our claim that some ETI civilizations may well be sufficiently long-lived to become IMETI to undermine the justification for SETI itself. In fact, that justification is strengthened. As one increases the average lifetime of ETI civilizations to allow them to develop an interstellar travel capability, then, barring interference between the civilizations, one also increases the number that can be expected to exist in the Galaxy at any particular time. In terms of the Drake equation, if we argue for a large value of L to give us IMETI, we also increase the total population of ETI accessible to SETI. However, one other interesting possibility is raised when we consider the possibility of an intelligent civilization. While radio might be the best means of long-range communication that we can envisage, this may not be the case for a civilization millions of years older than ourselves. This possibility belongs to the same realm as the notion of faster-than-light travel. Special relativity forbids it, but perhaps there is some bizarre physics we do not yet understand. Perhaps ETI has mastered communication technologies we have not even contemplated since we do not yet have all the physics we need. It is conceivable that an IMETI has set up broadcasting beacons, but not using "antique" radio technology. Perhaps they have neutrino beacons (*neutrinos* are strange "will-o'-the-wisp" subatomic particles that travel with the speed of light), or some other capability it is difficult to even speculate about. Of course, until we have the appropriate technology, we cannot read the signal. This is one of the plausible explanations for the failure of SETI to detect signals; our radio technology is simply too "old" for a highly advanced ETI.

Even apart from the possibility of advanced beacons, there will be other differences between the way that highly advanced ETI behave and the expectations we have derived from our own experience. It is possible that the home solar system of an intelligent species that has survived for a hundred million years or longer than us will be distinctly different from one of a younger species (or one that is uninhabited). This is due to the possibility that ETI's advanced engineering could enable them to modify their parent star and its attendant planets. The former possibility we will call *stellar engineering*, while the latter is customarily referred to as *terraforming*. Terraforming refers to the notion of bringing a dead planet to life. This would require a scientific understanding and technologies that are considerably beyond what we have at present. However, we can see such understanding and technologies as a natural continuation of even our existing technologies. If humans are to avoid an ultimate environmental crisis, then understanding exactly what can and cannot be done to planet Earth, and ways of ameliorating our worst excesses, must be found. These technologies are the stepping stones on the way to being able to create and regulate a living biosphere on a dead planet. (Of course, the object need not actually be a planet in the strict sense. Many planets in a solar system may well be gas giants, completely unsuitable for settling. However, these planets may well have rocky moons suitable for settlement. Jupiter is such an example in our Solar System.)

How would one spot the presence of terraforming at interstellar distances? In exactly the same way as one detects the presence of natural living planets at great distances: by observing the spectra of their atmospheres. Our present telescopes are not capable of analyzing the spectra from planets around distant stars, but that capability will surely come in the foreseeable future. How might terraformed systems differ from natural ones? Until we have actually detected a few natural systems, it is hard to know what is typical of them. However, the probability of the natural occurrence of two or more living planets in a single system is likely to be low. Hence, the occurrence in a planetary system of two objects that reveal the presence of living atmospheres, with their telltale atmospheric signatures, might suggest that one of them has been terraformed. At least this possibility would need to be considered, and it could be a tantalizing signal of advanced ETI (or alternatively a startling sign that the natural creation of living worlds is much more likely than we think at present).

Learning to modify a planet would be an awesome technology, yet more advanced societies might be able to modify the properties of their star. The modification of a star might be done in two ways. The first is direct modification of the star itself. The second is by encasing it in a Dyson sphere, as described in Chapter 1. Engineering on this scale is the mark of what Kardashev labeled type II civilizations, those that can handle energy of the order of the total output of a star.

The direct modification of a star is so far from our technological horizons that it is not straightforward as to why any type II or III civilization might want to do it. There are, however, four conspicuous motivations. The first, and most often highlighted, concerns the relationship between the energy consumption and lifetime of the star that we described in Chapter 1. While our Sun might still have billions of years ahead of it, it is conceivable that a slightly more massive star could both be suitable for the existence of living planets and also have a lifetime that is measured in hundreds of millions rather than billions of years. (Complex life-forms would have had to be sparked much earlier on the planets of such short-lived stars than on Earth.) To a long-lived ETI civilization that had already survived for a significant fraction of the expected lifetime of its star, the limited time span for the survival of its home planet would be deeply concerning and it might well want to do something about it. However, we have already played down the likelihood of the dying-star hypothesis when we considered the issue of interstellar migration.

A much more likely motivation for a first bit of stellar engineering might be stability against stellar flares. We described these vast eruptions from the surface of the Sun in Chapter 4 in the context of mass extinctions. While, as described in the previous chapter, we believe intelligent life will be reasonably robust against the threat of such total extinctions, there are still plenty of good reasons for an advanced ETI civilization to try to minimize the risk of them. Intense radiation bursts are associated with these eruptions. The threat to the inhabitants of any vast inhabited space structures outside the protective embrace of the home planet's magnetic field (such as a Dyson sphere) would be serious. We do not yet fully understand what causes the instabilities in the solar atmosphere that lead to solar flares. While we have a good idea of the basic nuclear physics that all stars have in common, we are only just beginning to appreciate their individuality. We have already mentioned the desire to study these stellar idiosyncrasies as a likely motivation for the first interstellar probes. If an understanding of the causes of the dangerous stellar flares can be arrived at, then trying to control them by modifying the star in some way may well be a worthwhile endeavor. Whether the modifications required to provide protection against flares would be observable at interstellar distances is hard to say, but it is a possibility worth keeping in mind as astronomers study stars in unprecedented detail.

We need to emphasize that the whole concept of stellar engineering is extremely speculative. However, if the idea of type II and III civilizations is to be invoked in support of SETI, then it is possible to speculate about the technological capabilities of such supercivilizations. Stellar engineering may seem quite beyond anything we could imagine at present, and it would be easy to dismiss it as no more than science fiction. However, what is inconceivable to us could be commonplace to type II supercivilizations, and quaintly naive to type III ultra-superciv-ilizations. If we are going to have the courage to embrace the intellectual concept of type II and III civilizations, then we must suspend the normal limitations on our imagination of the possible extent of advanced technologies. Stellar engineering and terraforming therefore remain on the "possibles" list for advanced ETI.

Observability at interstellar distances is not a problem for the third kind of possible modification, which is expressly concerned with communicating. While radio beacons are one means of sending an electromagnetic signal that can travel over great distances, to an ETI with a stellar engineering capability there is an interesting alternative. Rather than use energy running a beacon, why not make use of the unused energy that their star radiates off into space? If distinctive modifications could be made to the features of the star, an alternative form of beacon could be created. Exactly how such a stellar beacon could be made the distinctive mark of intelligence remains a complete unknown. But again we must allow our imagination the escape of appealing to the exceptional technological capabilities of ETI in a type II or III civilization.

The fourth kind of modification suggested has nothing to do with wanting to modify the star at all. Instead, it is a proposed means for advanced ETI to get rid of the large quantities of nuclear waste that will almost certainly start to be accumulated by a type II civilization (assuming that it is likely to have mastered the totally safe use of nuclear energy). Even if ETI decides that nuclear technology is too dangerous to be used for the prosaic business of generating energy on the home planet or guarding against one's enemies, it will almost certainly need to use it for certain space operations and all sorts of miscellaneous uses such as medicine and scientific research. On Earth it has already been seriously proposed to launch rockets to transport the planet's nuclear waste safely into space. If ETI is already undertaking nuclear dumping into its star, we might be able to detect some spectral signature of this form of garbage disposal.

The other way we noted in which an advanced civilization might modify the signature of its star is by the construction of a Dyson sphere. Remaining in the plane in which the planets go around a star limits the amount of radiation emitted from the star that is incident on surfaces where it can be converted into useful energy using solar power stations. If a civilization wishes to ensure that it can absorb very much more energy, as must a type II civilization, then it is suggested that it should enclose its star in a vast bubble of matter. Building a sphere (perhaps in the form of clustered space colonies) would require engineering on a quite literally astronomical scale. Much of the matter in the star system, whole planets and asteroid belts, would need to be utilized. Such a project would take many millennia of organized labor, but millennia are something an advanced ETI would not be short of. The benefit of having near-limitless energy could be worth the tremendous effort. Skeptics would be well advised to remember that if a long-lived ETI species is to allow its population to expand indefinitely, it will almost certainly need to build (and supply with energy) vast space-based settlements. The Dyson sphere concept would be the natural extension of this process. Even if the full sphere is judged an unacceptable effort, the construction of a *Dyson ring* or *Dyson cap*, covering only a smaller patch of the sphere, would be worthwhile (and could later be expanded to the full sphere).

Of course, if all or a substantial proportion of a star's radiated energy were being absorbed by the vast solar panels of a Dyson sphere-style structure, then waste heat from the structure would also have to be radiated and could be detected in the infrared region. Hence, a Dyson sphere would be visible as an object emitting a similar quantity of energy as a star, but with the emission in the infrared enhanced dramatically. If the structure was not a complete sphere, then some periodic variation in this spectrum might well be apparent, oscillating between the infrared and visible as the star rotated. Dyson proposed the possibility of his spheres in the early 1960s, in the comparatively early days of SETI. It has been a possibility at the back of astronomers' minds for many years, but as of now no candidate signals have been detected by any of the infrared surveys of the heavens. However, infrared astronomy using cooled space-based telescopes is still in its infancy, so the possibility of the future detection of a Dyson sphere certainly cannot be ruled out.

Interesting though the possibilities of stellar engineering and terraforming are, they do not require any new distinctive search techniques. Rather they might crop up in the course of standard astronomical investigation. When it comes to detecting IMETI at home, it seems that standard astronomical investigations including SETI by radio, time, and an open mind are all that is required.

IMETI on the Way

With the distances between individual stars being several light-years, at the travel speeds we have been thinking about, of the order of one-tenth the speed of light, we must think of many tens of years of travel time between even the closest stars. (We must stress again that speeds a tenth the speed of light are way beyond the present technology.) There are 400 billion stars in the Milky Way Galaxy. Given this, the most likely possibility is that even if IMETI exists, and decided to make use of its interstellar travel capability (probably as a robot craft), then it is not here because it is still on the way.

If there is any issue that allows a defense of the compatibility of the notions that interstellar travel is possible, and that IMETI will not be present in the Solar System, it is the question of expansion time scales. These determine how long it takes IMETI to get from there to here. It is obvious that this depends on factors whose values we cannot know, most particularly how far the IMETI's home planet is from our Sun. It is usual to generalize out the factor of the distance between the two home planets and be concerned simply with how quickly the IMETI spread through the Milky Way Galaxy. From these average times to transit Galactic distances and an assumption of random distributions of the origins of IMETI in time and space, we can calculate the chances of IMETI arriving in various time intervals. However, given the uncertainties in the calculation, what is generally seen as significant is the general magnitude of the time it takes to cross the Galaxy. Remember that it seems plausible that IMETI could be many hundred millions (or even billions) of years older than ourselves. Therefore, it would seem that the expansion time scales to cross the Galaxy must be less than tens or hundreds of millions of years if we are to expect a reasonable probability of IMETI actually getting here. If expansion time scales to cross the Galaxy are greater than 10 billion years (the age of the Galaxy), then IMETI is likely to still be exploring their "local" environment and, unless they are extremely abundant, they are not necessarily going to be getting anywhere near us.

So then, what are the best guesses for these expansion time scales? Actually many scientists have believed they can offer rather more exact answers than just best guesses. They use mathematical models to calculate quite precise measures of expansion time. They are able to do this because random diffusion processes are something that physicists have studied in great deal. If one can specify accurately the parameters involved, then the actual modeling is quite simple. One popular form of modeling, the so-called Monte Carlo method, works by sending a large number of random tests through the system one after another, and then calculating the average result. Anyone with a flair for writing computer programs can do this sort of modeling on a home computer. However, what values of parameters should one specify?

Some of the numbers, such as the average speed of an interstellar spaceship in flight, we cannot know, although a plausible guess of a tenth of the speed of light is sensible. A more serious problem concerns how we calculate numbers that determine the rate at which IMETI set out for new star systems, how long they stay there, and after having been there, whether they strike out for a new star farther away from home. Many studies assume that this is the case because the situation they model is colonization driven by increasing population. We have a reasonable idea of how situations like this work from archaeological studies of the spread of our own species, for example, across the islands of the Pacific from Asia, and down the continent of North America from the far north. Even in these cases, the specification of numerical values is rather arbitrary. There will be only limited resemblance between the movements of primitive hunter-gatherers largely motivated by the search for new sources of food, and the migrations of extremely high-technology IMETI whose motivations we are uncertain of. Thus, it is not helpful to push the human colonization model too far, despite the obvious attractions to science-fiction writers. We have already argued that we do not believe that the initial motivation for interstellar travel can be colonization of this sort. We have surmised that IMETI will not be seeking living space, but knowledge and adventure. Given this, and the fact that they will almost certainly have instruments sensitive enough to obtain spectra off the planets of nearby systems, the idea of portraying their progress as random is surely inaccurate. Rather they will move between systems that show significant signs of being interesting, such as those with living planets, or unusual varieties of stars. IMETI will surely not be hunter-gatherers.

Some researchers have argued that since it would take just a million years to cross the Galaxy at a tenth the speed of light, and even allowing for a very generous dwell time for any ETI colonists on one planetary system before they send a new generation of their colonists outward and onward to new colonies, it could take just a few million years to colonize the Galaxy. This is the simplest form of the argument, If they are there, why aren't they here? This approach is based on the assumption that IMETI is motivated purely by a colonization instinct. But why should this possibly be the case?

If the expansion of IMETI is driven by factors other than the aggressive acquisition of new territory, then the idea of describing it as random diffusion breaks down in several ways. First, there will not be a random choice of which star to visit next. Second, each new mission might be initiated from the home planet, rather than from the perimeter of the colonial empire. This would slow down substantially the journey to the next star out, because it becomes necessary to travel again the distance to the last star reached. Third and perhaps most importantly, there is no longer clear motivation for indefinite expansion. Only so many different types of stars are worth taking a close look at, and only so many alien biospheres are worth cataloguing. Eventually, a scientifically minded IMETI will "know it all" or "know enough," which somewhat undercuts knowledge acquisition as a motivation for continuous expansion. The idea that IMETI might just explore the area of the Galaxy nearest to it, and then stop, contented with its lot, rather than necessarily expanding indefinitely to engulf the entire Galaxy, significantly weakens the view that the existence of long-lived IMETI necessitates that our own locale be explored.

So might IMETI still be on its way here? We can merely confess our ignorance and say that yes, it might; and then again, it might not. Certainly it seems unlikely that there was an explosive colonization of the whole Galaxy in, say, 100 million years that would make it almost certain that IMETI would have been here by now. Visitation of Earth by IMETI in either the distant past or the future is a possibility, but by no means a certainty, even if the abundance of ETI is toward the upper limit of our expectation of, say, a few thousand in the Milky Way. Of course, without being too immodest, we think we can expect that our blue-green planet with its clear evidence of life would be something of a target for any scientifically minded ETI (or even one looking for additional habitats!). How much this tips the scales in favor of a visitation is open to debate. But the prospect of there being IMETI is something SETI needs to include in future search strategies. It is no longer appropriate for SETI to turn a blind eye to the Fermi question. It is a legitimate scientific question, demanding serious scientific consideration, and if the SETI exponents wish to convince us with their arguments in favor of ETI, they must accept that those same arguments can be extended to IMETI.

If IMETI is on the way here, is there anything we can do to try to detect it? Would interstellar spaceships, scorching across the Galaxy driven by nuclear propulsion, leave any traces that could be seen at interstellar distances? The consensus is that there is surprisingly little chance of detecting interstellar vehicles at any substantial distances. One notable exception has been proposed by the imaginative Freeman Dyson. Remember that any interstellar ship, as well as needing to accelerate at the beginning of the journey, will also need to slow down at the end. The most obvious braking mechanism is simply a nuclear rocket pointed away from the direction of travel, essentially similar to that used to accelerate.

Dyson proposed, however, that this might be replaced, or supplemented, by brakes based on the interaction between the spaceship and the extremely thin gas to be found in the interstellar medium. This is possible because the gas is in the form of ionized plasma, and hence there is an electromagnetic interaction between the plasma and any conducting material moving through it. This is known as an *Alfven engine* and has led to some satellites in the Earth's ionosphere experiencing a much higher drag than had been expected. Producing the relevant braking effect from interstellar speeds would require vast networks of extremely thin conducting wires trailing behind the spaceship for years as it very gradually transfers the vast amount of energy bound up in its motion to heating up the interstellar medium. This would leave huge hot "skid marks" of plasma across space. Hot plasma cools by emitting radio waves that can then potentially be detected by Earth-based radio telescopes. The traces would have the extremely distinctive feature of being straight for the vast majority of their length. If long narrow streaks of plasma are detected, not associated with any other obvious source, then perhaps these are the result of trying to slow down interstellar ships. Detection may seem unlikely, but the suggestion is very typical of the imaginative hypotheses put forward by Dyson. At least it gives some idea how the thing could be done, even if it seems rather improbable. It seems to us that detecting IMETI in transit is somewhat unlikely. But we might be proved wrong.

IMETI Come and Gone

It should be clear that we believe it is possible for IMETI to have developed in our Galaxy at any point in the past few billion years. Hence, at any point in the last couple of billion years it is possible that our Solar System, and even our own planet, was visited by IMETI. This gives rise to two conspicuous questions. First, is there any way we can tell whether IMETI has visited the Solar System or Earth in the distant past? Second, isn't it the case that, there being so much more of the past than there is of the immediate future, it is very much more likely that IMETI has visited us in the last billion years than that it is likely to do so in the next 200 years or so? If this is the case, then surely any resources that might be used looking for IMETI visiting in the immediate future would be far better dedicated to "extraterrestrial archaeology." We will address initially the argument behind the second question, and then show how, in the light of the best answer we can give to the first, it loses its force.

Let us compare the two stretches of time. The first stretches from the earliest time when ETI could have arisen in our Galaxy to the present. The second stretches from the present for, say, the next couple of centuries. This second period of just two centuries is chosen because it is about the maximum we can reasonably plan for, even at our most abstract. (The only people we can think of who have to plan for more than 200 years are those responsible for the disposal of high-level nuclear waste.) The first stretch of time is about 10 million times as long as the second. This means that if we assume that the chance per year of being visited by IMETI is constant, then we are 10 million times as likely to have been visited in the past as we are to be visited in the immediate future. Of course, the possible rate of visitation may not actually be constant. The chance of us being visited by an IMETI is approximately proportional to the number of IMETIs that exist in the Galaxy. Given that, as we argued in the previous chapter, advanced ETI and hence IMETI will tend to survive for long periods, this number of IMETIs in existence can be expected to have grown throughout the past 2 billion years. The rate of growth of the number of IMETIs may well even be increasing, depending on the "demographic" distribution of stars, and the length of time it takes to get from simple biological slime to space-traveling IMETI. However, even this will not be enough to balance out the total probabilities of arrival in the two lengths of time. Consider the most recent million years of the past. It is pretty unlikely, given the sort of abundances we expect, that more than one or two new IMETIs will have come into existence in that period, and most probably the number will have been constant. Hence, the chance per year, even considering the increase in IMETI numbers, will be common for the previous million years and the immediate future. Given that the previous million years is still much longer than the future time frame we are considering, it is nearly certain that were an IMETI to arrive in the period including the previous million years and the next few hundred years, then it will have happened in the first section of that period rather than the much shorter second section.

Thus, we can be fairly certain that the increase in the number of IMETIs in the Galaxy will make no difference to our calculation that any extraterrestrial visitors are much more likely to have turned up in the past than to turn up in the foreseeable future. A consideration that is less easy to gloss over is the possibility that the IMETI's motivation for visiting us will have increased sharply in the comparatively recent past. If, as we have already argued, the main motivation for interstellar travel is research, then any changes visible in our planet from interstellar distances will be an enticement for visitation. We judge that the Earth has sent out two spectacular signals that would have been apparent to any IMETI monitoring the planet fairly carefully. The first is the change in the composition of our atmosphere with the origins of photosynthesizing plants, which saw the levels of oxygen increase dramatically. This could be detected by the distinctive presence of certain spectral emissions, for instance, that for ozone, in the planet's spectrum. This would have occurred some 2 billion years ago, and would have indicated to any observing ETI that here was a life-bearing planet. The second is the sudden burst of radio radiation associated with the dramatic growth in radio and television broadcasts that started about 50 to 60 years ago. The expanding spherical front of Earth's radio and television broadcasting will now be reaching stars out to a distance of 50 to 60 light-years, becoming fainter with distance in accordance with the inverse square law. We discussed in Part A how this would probably be observable, not so much because of the strength of the signal, but because of the curious periodicity it would show as the major sources of broadcasts, North America, Europe, and

Japan, rise and set with the rotation of the Earth.

It is important to remember that there must be time for the signal from Earth to travel to wherever the IMETIs are and then for them to travel here. In the case of the first signal, there is no problem. The Earth has been conspicuously biologically interesting for anyone who can detect a spectrum from it for almost all of the past stretch of time that we are interested in. The signal based on radio broadcasts is a different matter, as it is a very recent phenomenon and a very conspicuous trigger for extraterrestrial visitation. (It is also a conspicuous trigger for targeting the Earth with the sort of beacons that SETI by radio looks for. This was the scenario described in the movie *Contact*, where the first signal received was a returned version of the first television signal sent from Earth, Hitler opening the 1936 Olympics.) Radio signals might well generate a substantial increase in the chance per year of visitation, if IMETI lies within the distance such that the signal can reach them and they can reach us in the available time frame. That distance in light-years will be approximately the number of years since the mass use of radio began divided by about 11 (1 year for the signal to travel at light speed plus approximately 10 years for a spaceship to travel back at a tenth the speed of light). Hence, a bubble about 20 light-years in radius is defined during which our radio signals are received and any IMETI that had set out in search of the originators of the signals would have time to reach us during the next 200 years, the future period we are concentrating on. If there are any IMETI within this bubble (which is not very likely, in terms of the incidence of ETI we argued for in Parts A and B), then we could get a visit in response to our radio calling card during the next 200 years. Of course, this argument supposes that IMETI could detect our radio signals and then visit. Is it not more likely that they would first try to establish contact by radio?

Despite the significance of the increase in the chance of visitation due to the Earth's starting to broadcast, it still seems unlikely to offset the vast difference in size between the past and the planned-for future. After all, remember that the Earth's biological nature could already have been a substantial motivation for visitation over the last billion years or so, and that signal will have been available to any distant IMETI who can sight the Earth with its giant telescopes. Still, the existence of the radio trigger does assure us that we are not engaging in completely ridiculous "presentism" in imagining that ours may be the era in which IMETI first arrives.

If it is more probable that IMETI arrived in the past than it is that IMETI will arrive in the near future, is it not the case that we should concentrate our efforts on searching for traces of past visits, rather than present or immediate future arrivals? All things being equal, undoubtedly we should. What needs to be equal is the availability of a straightforward methodology for detecting what is being searched for. In the case of SETI by radio, this is clearly available. We will discuss searching for present arrivals shortly. However, searching for previous visits to our Solar System strikes us as being very challenging indeed.

The most obvious method is to look at recorded human history. This approach is popular with a certain community who claims that Aztec paintings or Mesoamerican pyramids show some image that evokes a spaceship or an alien spaceman. Such claims strike us as somewhat bizarre. Exactly the same argument used earlier based on the enormity of the past can be made, using recorded human history (rather than the foreseeable future) as the smaller period. Moreover, nothing associated with the onset of human civilization would be observable from any astronomical distance. Extremely strong evidence, rather than vague hints, is needed to confirm a hypothesis of IMETI visitation during recorded human history.

If we cannot rely on even indirect historical reports, then we will have to rely on physical evidence, left either deliberately or inadvertently by visiting IMETI. It seems entirely plausible to us that if the primitive Earth were visited by IMETI, it might well have gone to the trouble of leaving some indication of its passing, either for some intelligence that might arise on the planet or for another passing IMETI. However, it is not easy to build a monument that both is obviously the product of intelligence and can last for hundreds of millions of years, which is the time scale that any monument would have to survive. Certainly, no such monument could survive on Earth. Any traces of its original form would almost certainly be eroded by the action of atmosphere, microbes, and geological forces. The two obvious locations for a survivable monument would be either on one of the rocky bodies of the Solar System that lack an atmosphere or in orbit somewhere. The latter, however, would not be very effective as a marker, because unless it was radio emitting, it would be very hard to find, and it is hard to imagine designing a radio transmitter and power source that could keep going for the relevant stretches of time. (The "invisible monument" idea was used by NASA when plaques with engraved coded messages were placed on deep-space probes as a form of interstellar "message in a bottle." The prospects of them ever being found by ETI in the depths of the cosmos are so remote as to have made the gesture purely symbolic.)

Thus, the most likely form of monument would seem to be some sort of construction, hopefully visible from orbit, on one of the rocky bodies of the Solar System. Some thought they had identified such a monument in the face-like feature observed by the Viking probe in the Cydonia region of Mars. However, human-like faces are a most unlikely form of monument (as we argued in the previous chapter, ETI certainly will not be "humanoid"). In addition, we humans, because of the setup of our visual system, have a tendency to see apparent "faces" in boundaries differentiating patches of light and dark. (Spend a minute or two gazing at the clouds, and note how many "faces" you see.) Recent high-resolution investigations of Mars revealed that this is the cause of the "face" on Mars, rather than any genuine face-like surface feature. Similar factors accounted for the earlier sightings of alleged canals on Mars in the early years of the twentieth century. The most plausible form of monument to intelligence would be some kind of geometrical figure, for instance, the geometrical representation of the Pythagorean theorem. However, until high-school geometry exercises are observed carved on Mars, or one of the more stable moons of Jupiter, we have to assume that if IMETI has visited the Solar System, it left no lasting monument to its transient presence.

Would IMETI leave inadvertent traces of its passing? Some,

most notably the physicist Frank Tipler, argue that these "traces" would be spectacular. Indeed, Tipler believes that had IMETI visited the Solar System with robot probes, it would have torn it apart in the search for materials to build more probes to send to other solar systems. These theoretical self-replicating probes, known as *von Neumann probes*, would rapidly fill the Galaxy, constructing an exponentially increasing number of themselves. However, we believe that such Galactic vandalism is an extremely unlikely motivation for interstellar travel. Similarly, we have already argued that colonization (even by robot probes) is unlikely to be a motivation, so we should not look for traces of IMETI colonies. Instead, it is more likely that we have to look for the meager markers left by a scientific mission, akin to finding a discarded theodolite deep in the ice in Antarctica from a pioneering scientific mapping mission.

Objects such as instruments left on our planet are no more likely to have survived than deliberate monuments. Objects left either in orbit or on the rocky bodies of our Solar System might well have survived, but they would be very very hard to find. One way of looking for the presence of probes might be to try to detect the radioactive tritium that would be left by a nuclear rocket based on the fusion of hydrogen. Such a search has been proposed by Michael Papagiannis, the chairman of the International Astronomical Union's Commission on Bioastronomy. Unfortunately, the half-life of tritium is only 12 years, meaning that any traces of "exhaust fumes" left by a nuclear rocket would have declined to a tiny fraction of their initial density over just a few centuries. Only presently active probes are likely to be detected by this method. Searching for presently active probes takes us on to our next topic. However, let us pause to summarize our thoughts surrounding the possibility of detecting IMETI that previously visited the Solar System.

Although the long history of Earth makes it more likely that IMETI visited in the past than that it will happen in the near future, this does not mean that we should concentrate on the search for past probes to the exclusion of the present. For a start, the possibility that our planet's recent mass use of radio communications will have acted as a trigger to IMETI visitation suggests that we are living through a period where the probability of visitation may be much higher than it has been historically. Even if IMETI visited our Solar System in the past, it is hard to guess how we might come to know about it. Certainly, there is nothing even vaguely like the well worked-out methodology of SETI by radio. Hence, despite the fact that IMETI more likely visited us at some point in the past, it is more probable that we will detect the present or future presence of probes than the traces of ancient IMETI.

IMETI Here and Now

We have pretty much given up hope of detecting IMETI either on its way here, or by virtue of any traces it might have left on a previous visit. This leaves us with two possibilities: detecting it at home, by virtue of SETI by radio or the projects discussed earlier in this chapter, or detecting the presence of IMETI spaceships in our Solar System here and now. It must be pointed out that if there are uncertainties as to whether there is ETI anywhere, these are multiplied a millionfold with regard to the possible presence of IMETI in our Solar System now. We can have no more assurance than that it is a genuine scientific possibility. However, as was the case with the ancient probes, genuine possibility is no guarantee that we have any means of finding out whether that possibility is realized. How might one search for either the present existence or arrival in our Solar System of interstellar spaceships created by an alien intelligence? We are about to move up an octave on the scale of speculation, and no doubt on the scale of incredulity of the skeptical reader. We apologize for that, but did warn at the commencement of Part C that we were prepared to confront the giggle factor head on.

To some, the idea that we would need to look for the arrival of an IMETI visitation seems foolish. They argue that it would be among the most conspicuous events in human history. IMETI, they argue, might be friendly or malign, but it would be very curious of IMETI to come all this way and then not draw attention to itself in a spectacular fashion. However, to us this speculation into extraterrestrial psychology, that most conjectural of disciplines, seems misguided. True, the most obvious analogies we have in Earth history to such an event, for example, the arrival of the Spanish in America, rarely involved the overuse of discretion on the part of the visitors. However, the motivations of an IMETI are unlikely to center around that unusual combination of religious conversion, colonization, and economic pillage that has typified the first contact between long-separated human cultures. Indeed, as we have already argued, study and knowledge would likely be its principal motivation, and it might well be in its interests to avoid overly disturbing the system under study. Add to this the fact that any probe need not necessarily be a very grand affair, but might be an uninhabited highly instrumented robot device, and the notion of the grand arrival seems less likely. It might well be that we are going to have to look extremely hard to detect our "visitors."

We have already mentioned the most scientifically respectable proposal for probe search, which involves searching for the tritium that would be associated with any fusion-powered source of propulsion. Another proposal is to hunt for gamma-ray emissions associated with a nuclear-powered probe. This is technically feasible, using instruments on satellites that were put into orbit to monitor for atmospheric nuclear tests and also with space detectors designed for gamma-ray astronomy. Another approach for the future is to look for the gravitational waves that might be emitted by some of the more exotic propulsion systems that have been proposed; however, this would require detection technology with a sensitivity not yet achieved for gravitational wave astronomy. What these proposed techniques have in common is that they involve searching for alien probes using accepted scientific techniques. They thus have the advantage that were an object to be observed in our Solar System whose characteristics could be understood fairly clearly as those of some propulsion system, this evidence should be pretty robust and probably rapidly become widely accepted. The disadvantage of these techniques is that they can only detect a small subset of possible probes.

These difficulties with more physics-based approaches are

why ufology, the most famous means of searching for IMETI probes in our Solar System, still has a place. The systematic study of the numerous sightings of high-altitude lights or unrecognized vehicles might well be the best way of detecting any IMETI probes. If the chosen strategy of IMETI visitors were simply unobtrusive research into our biosphere, then occasional sightings of its exploratory landings might be the most we could expect. These are the sorts of things that will be witnessed by those on the scene, rather than those with the appropriate instruments. Such a form of research, where the main "measuring apparatus" is in fact a nonscientifically qualified population, is not unprecedented. Considerable information about the ecology of some of the Earth's densest habitats, such as tropical rain forests, has been gained through "ethnobiology," where the inhabitants of a region are interviewed by appropriately skilled biologists. Such migration of knowledge from the populace to the scientific elite has even been known in physics, although it is very much the exception to the rule. The classic case is that of meteorites, which were reported by many ordinary people while completely excluded by the scientific establishment. For example, on September 13, 1765, people in fields near Luce, France, saw a stone mass drop from the sky after a violent thunderclap. In a report to the Academy of Science, Lavoisier, now regarded as the father of modern chemistry, made it clear that these people must have been mistaken. It was not until 1803 that the Academy would accept the reality of meteorites.

Of course, a scientific investigation based on the reports of a mass populace is fraught with difficulties. One of the very few scientists who persisted with the investigation of UFOs after the Condon report was J. Allen Hynek, who had previously worked on the subject for the U.S. Air Force. His classification system for reported sightings was responsible for the title of the movie *Close Encounters of the Third Kind.* He described the task involved as follows:

> What we have here is a signal-to-noise problem: There is indeed a fantastic amount of noise, represented by the many misidentifications of familiar objects seen under unusual or sur-

prising circumstances—balloons, birds, satellites, meteors, aircraft, stars—yet, in all scientific honesty, one is led to ask
whether there might not indeed be a signal somewhere in the
noise.

Whether deliberately or not, Hynek's language was strangely
evocative of that used by the radio astronomers who practice
SETI. Remember Morrison's call to the radio astronomy community to search the wave bands and the ray direction "down
into the noise." Hynek was right that there is all too much
"noise" in claimed UFO sightings, and the near-mass hysteria to
which certain factions of the American population have been
driven by supposed alien abductions has not done anything to
reduce it. Hynek was right to assert that there is probably a signal lurking behind the noise. In fact, there might well be several
signals there, by no means all of which have anything to do with
ETI.

Among the less relevant forms of signals are the sightings that
can reasonably be attributed to a range of classified aircraft
operated by the American military. The existence of these has
been persistently denied by the Pentagon. However, the undeni-
able existence of previous programs, now declassified, and the
extraordinary levels of security surrounding the Nevada facilities
where this equipment is usually tested (including the infamous
Area 51), strongly suggest otherwise. Of course, those who like
to link their aliens with conspiracy theories believe that the
IMETI "signal" and the Air Force "signal" are one and the
same. To us, however, the activities of the military, and the pecu-
liarly tight secrecy that they are covered by, are a distraction
from the serious business of looking for IMETI.

It seems very likely that many other claimed UFO sightings
can be explained by a range of high-altitude weather phenomena not yet fully understood by science. Many UFO sightings are
of randomly moving points of light at a high altitude. It might
be that this is some high-altitude analogue of what at surface
level is called *ball lightning*. These tight balls of electrical energy
display strangely purposeful activity as they follow the invisible
lines of electric field in the environment. At high altitude they

might well display the sort of high-speed formation movement that has been reported by many responsible witnesses. Strictly speaking, a science of ufology includes the study of such meteorological phenomena. However, when we discuss here a scientific ufology, we mean the more intriguing business of possible IMETI detection by the study of the UFO phenomenon.

How might one possibly confirm the existence of an IMETI probe by ufology? Ultimately what is needed is clear evidence on three points: that the proposed object was real, and not an optical illusion, something seen in a dream, or an ordinary object seen in an unusual circumstance; that the object is the product of intelligent design and not a natural physical phenomenon; and that the object originates from outside our Solar System.

Obviously evidence on the three points is connected. Probably the hardest condition to meet is the third, although we can think of several ways that would be sufficient. The most obvious would be repeated and well-confirmed observations of an object clearly not constructed in our Solar System (such as if a 20,000-ton spaceship was found on one of the moons of Jupiter). In addition, one might make use of some of the properties of ETI that we have described as following from our accepted theories. Examples would be the finding of biological material with a biochemistry other than that found on Earth, or clear evidence of the use of some nuclear rocket technology. Ideally, any evidence would also provide a basis for discovering more information, thereby allowing us to increase our confidence in a claim of discovery. These are not easy requirements to meet. Nor should they be. The object of our scientific ufology is not to demonstrate that IMETI probes are present, but to offer as good a chance as possible that we will come to a firm and deep knowledge of them if they are present. If we make our hurdles too low, then our chances of error increase sharply.

What investigation would be able to meet these standards? There could be three possible strands to this scientific ufology, each with its own character, although all are closely related. The first would be to use special wide-field telescopes to observe fast-moving objects in the sky. The second would be the most like the forensic approach described in the previous chapter or the original Condon report, in that it would concern itself with popular reports of UFOs in general, albeit from a robustly scientific viewpoint. The third would concentrate on examining in detail only the most tantalizing reports thrown up by the broader searches. Through a careful blending of these approaches, we believe it would be possible to conduct an investigation that quite likely could identify any IMETI signal within the UFO noise. Let us describe in a little more detail what might be involved in each project.

The key to the first approach is to have a network of wide-field telescopes looking at the sky. In Chapter 4 we mentioned the Spaceguard System proposed for detecting near-Earth objects (asteroids and comets). These wide-field telescopes would pick out the movement of fast-moving objects. There is an additional existing capability with the military telescopes that monitor the motion of satellites across the sky. With Spaceguard or military telescopes, detection can be achieved by simple filtering. What moves too fast is an airplane; what moves with the rotation of the Earth is a star. What is left moving swiftly across the sky, but not too swiftly, is the bright point of light reflected by a satellite in orbit, for example. If one looks up at a dark night sky, not overly illuminated by the lights of a nearby city, with the naked eye one can see reflected light from these moving satellites. What a wide-field telescope network such as devised for Spaceguard offers is the chance to automate the tiresome business of having to watch and catalogue every object in the sky. Near-Earth objects would be seen moving against the background of stars, but slower than satellites. Given current concerns about the dangers posed to human civilization by the impact of an asteroid or a comet with the Earth, a specialist network for spotting these objects is needed, without being compromised by military requirements. Given that the network would already be tracking bright objects in the night sky and calculating their speed and path, the Spaceguard System would seem ideal for a bit of part-time UFO hunting. If the network could automatically provide data on anomalous moving objects (for example, those that do not follow straight paths or display variable speeds), they could be studied in more detail. This would be particularly interesting if the observations confirmed sightings already reported by other means. Equally, such clearly quantified data on paths might be the first step to understanding the high-altitude meteorological phenomena that we earlier postulated might be responsible for a particular class of UFO sightings.

On this proposed approach, what would count as definite evidence of an IMETI probe? Obviously, a single sighting of a light source that is just outside the conventional bands is not good enough. However, the observation of a small but distinctive class of anomalous objects, along with the dates and times of their appearances, could be very suggestive. If it were possible to trace these appearances to a particular origin, or to show that they had a tendency to favor certain areas, this information could be fed into more detailed searches. While this approach cannot meet our three conditions on its own, it can certainly provide strong evidence on the first condition, the reality of the object observed, and provide cases that can feed into other forms of investigation. We might even spot a few new asteroids while we're at it!

While the optical approach closely resembles several standard forms of astronomical observation, the forensic approach is very different. Very few scientific disciplines require their practitioners to deal with confused reports from lay observers, these reports often being of dubious veracity or made by individuals who are mentally unstable. This will be the inevitable fate of those in the front line of trying to sift out the genuinely interesting UFO sightings from the mass of totally spurious reports. Practitioners will require a vast range of specializations, a wide knowledge of disciplines ranging from witness psychology to forensic photography to meteorology. Also essential will be a full knowledge of the properties of the various electronic devices that are used to record UFO data. Particularly important will be experts in the difficulties of authenticating radar signals. Many of the most persuasive UFO sightings are those where there is a simultaneous radar and visual report of an unidentified object. However, radar is a notoriously complicated technology, prone to false signals due to various forms of atmospheric distortion or electromagnetic interference. Because these false signals can affect critical uses of radar, such as for air traffic control, for safety purposes the target aircraft are required to carry transponders to assist the instruments on the ground.

There can be no doubt that this aspect of ufology will be extremely challenging, but a few previous endeavors do give one hope that from all the noise, a small signal might be extracted. Most noteworthy among these is the work of a small subgroup of the French national space agency CNES. This group, initially known as GEPAN, and later SEPRA, collects data from civilian authorities such as the police and civil aviation authority. It compiles records of their investigations, and follows up particularly interesting reports themselves. Their overwhelming conclusion is that the number of events that remain unexplained after careful analysis are neither numerous nor frequent. Of 3,000 reports, only about 100 required detailed investigation by the group, and only a handful of these have not been satisfactorily explained as natural phenomena.

There are two ways of interpreting this pattern, which replicates that found by the Condon report many years ago. The skeptical outlook would be to infer that since the overwhelming majority of reported events seem to be well understood by completely conventional means, even the remaining unexplained cases could also be so understood. The optimistic approach would be to say that this initial inquiry has acted as a filter and that the residue of cases are liable to reveal interesting novel phenomena, including potential IMETI. What decides between the two views? To our way of thinking, the optimistic approach is supported because of the reasoning that takes us from the premises of SETI to the possibility of IMETI probes in our Solar System. Against the background of this reasonable possibility, we believe the treatment of the initial data should be to filter out cases for further attention rather than to show that all cases are almost certainly capable of being resolved conventionally. Is this not the approach adopted by conventional SETI—excluding well-understood sources of interference, then concentrating attention on what is left?

The chances are that the investigation of the confused mass of UFO reports will not be overly rewarding. The day-to-day investigation will be difficult, and usually fruitless. Even the best-reported events will not be conclusive. It is likely that only real enthusiasts, or virulent skeptics, will be willing to dedicate substantial proportions of their careers to this work. But the uncertainties in the search, and its complexity, may be no greater than the uncertainties faced by those undertaking SETI by radio. It is to be hoped that the prior sentiments of the science ufologists might not overly color the nature of their conclusions. Certainly it need not. The history of science is full of cases where scientists have been forced to dismiss their pet theories, or have been converted to causes they have long opposed. Even when individual scientists have not been quite so rational, the net output of the community usually has compensated for individual idiosyncrasies. We definitely support the establishment of groups similar to the French SEPRA team. National or regional groups would be best, since an understanding of the local language, culture, and geography is highly desirable. (This conclusion is similar to that reached by a committee chaired by Professor Peter Sturrock of Stanford University, supported by the Society for Scientific Exploration, reported in mid-1998.)

What about the residue of cases that cannot be understood? The first issue to highlight is that these do not confirm that the sightings are definitely of IMETI. "We don't understand" does not establish anything, and appearances are often deceptive. Just as SETI cannot confirm the existence of ETI from unexplained signals that are not repeated, such as the Wow signal described in Chapter 2, so ufology cannot reach a positive conclusion from reports that cannot be understood. Reports need to be understood as likely sightings of IMETI probes if they are to have genuine evidential probity. How is this understanding to be reached? The answer to that depends on the nature of the report on a case-by-case basis. If possible ground traces of spaceships were to be discovered, for example, with attendant radioactivity, then expertise in radioactive contamination would be called for. Where radar data are crucial, then experts in that field would be needed. If strange metals were discovered, then metallurgists would be needed. If repeated sightings seemed to be associated with a geographical area, then appropriate instruments, such as automatic video cameras and magnetic anomaly detectors, should be set up on location. This has already been done at Hessdalen Valley in Norway, where various forms of unexplained light have been seen.

The important requirement here is that the individual events are likely to require experts from all sorts of fields to make occasional contributions. We suggest that such experts treat such requests as they would requests to cooperate in more standard scientific endeavors. Where possible, time and attention should be given willingly, and the usual scientific standards of open-mindedness and intellectual honesty should be upheld. In requesting this, we recognize the problems faced by scientists when confronted by the giggle factor. Conventional SETI disposed of the giggle factor, and there is no reason why an expanded SETI including a scientific ufology should not do the same. The investigation of UFOs might be compared to the investigation of the heavens. For many centuries, that investigation was dominated by the pseudoscience of astrology. Ultimately, however, an alternative approach proved to be the foundation of modern science in the development of celestial mechanics. Perhaps ufology can also cross the line, and become a robust scientific discipline.

Observation

We cannot say that there is more than a slight possibility that IMETI is in our Solar System or might arrive in the near future (in response to our signaling our presence in the growth of broadcasting). Even if it is here, there is no guarantee we will spot it. Even if we spot it, there is no guarantee we will be able to make the observation stick by developing an understanding of the episodes that confirms the sighting as being genuinely of ETI design. But there is a chance, and ultimately the spirit of inquiry is all about taking reasonable chances.

Epilogue: What Next?

"Dim and wonderful is the vision I have conjured up in my mind of life spreading slowly from this little seed-bed of the solar system throughout the inanimate vastness of sidereal space."

H. G. Wells, *War of the Worlds*

In this book, we have not attempted to disguise our enthusiasm for modern SETI, the search for extraterrestrial intelligence. It represents one of the great challenges for modern science, although sadly this fact seems to have escaped acceptance by many in the scientific community, the media, the public at large, and governments. Too many people have been overly dismissive of SETI, without looking at the evidence that supports it or thinking about the profound consequences of detecting intelligence elsewhere. We have tried to paint a realistic picture of the prospects for modern SETI in Part A; not too optimistic, not too pessimistic—just right!

Everything we know in science suggests that life in at least a simple form is likely to appear at any suitable location in the cosmos. Simple life is a natural consequence of simple chemistry (even if some of the details of that "simple" chemistry are not yet fully understood). Science has taken human understanding from an Aristotelian Earth-centered universe, to a Copernican Sun-centered universe, to our present awareness of a vast and violent universe within which the Earth is a mere speck of cosmic sand buffeted by the tides of universal change. The total insignificance of humans on the cosmic scale is now fully appreciated.

The basic building blocks for life are found everywhere. The chemistry of (water-based) life is reasonably well understood. Although the progression from amino acids to the replicating

magic of DNA remains uncertain, and may involve a hypothetical "RNA world," there is every prospect of a full understanding
of this progression being established in the laboratory in the foreseeable future. Planets around Sun-like stars are being found, and
with a new generation of space experiments Earth-like planets
displaying atmospheric signatures of life are sure to be discovered. Astrobiology (the science of the origin and evolution of life
in the cosmos) has a new legitimacy. Everything is in place, in scientific terms. The things that are now needed are more talented
scientists and engineers eager to pursue the astrobiology cause,
adequate funding, time, and a great deal of patience.

We need to be able to demonstrate that life out there is not
merely "slime." It appears that simple life on Earth formed very
quickly and easily once the primordial Earth had cooled sufficiently. One of the principal aims of the next generation of
probes to be sent to Mars is to see if simple biology formed
there as well. There is already tantalizing evidence that it did.
The progression from simple life to complex life on Earth was
erratic, and involved a sequence of freak occurrences. The
emergence of intelligence followed an even more tenuous path.
It is tempting to conclude that although simple life will form in
any suitable planetary environment, the emergence of intelligent beings elsewhere in the cosmos may be a rather rare
occurrence because of the necessary concatenation of unpredictable events. In Part B we tried to give an honest assessment
of the special conditions that led to the emergence of life on
Earth, and its tortuous path to intelligence. It was touch and go;
humans only just made it—and the future is full of cosmic haz-
ards and terrestrial uncertainties.

Life-forms in the cosmos may be aplenty, but all the evidence
from addressing the SETI, McCrea, and Fermi questions suggests that intelligent beings and advanced technological
civilizations may be relatively rare. That is the inescapable conclusion we have reached by looking at the issues from a variety
of perspectives.

It now appears that the early SETI experiments were based on
overly optimistic estimates for the abundance of ETI. The "cock-
eyed optimists" of the 1960s were talking about millions of ETI

civilizations in the Milky Way Galaxy, which if true would have meant a detection a week on average with the current generation of supersensitive SETI programs. It just was not going to be that easy. With our present understanding of astrophysics, the chaotic path to intelligence on Earth, and the negative results of over 30 years of SETI, a more realistic estimate is that ETI civilizations in the Milky Way might be restricted to a few thousand at most. Certainly an estimate of ETI in the thousands (rather than the hundreds of thousands or millions) remains compatible with the failure to make certain SETI contact after decades of search.

Based on theories no more secure than our understandings of the emergence of ETI, billions of dollars have been invested by governments in programs to hunt down elusive subatomic particles. The fact that such experiments have found the particles predicted by theory has justified the expenditure on research that is of no practical value but that nevertheless has increased human understanding. We do not wish to be overly critical of such investments; however, for a much more modest expenditure governments could have mounted successful SETI. But they lost their nerve. It seems that the popular "alien" culture peddled by the tabloid press and the pseudoscience fringe has counted against genuine SETI. Other areas of science have not had to compete with the giggle factor that was initially faced by SETI and would have to be overcome by a genuinely scientific ufology.

With BETA, SERENDIP, and Phoenix projects, the power lacking in earlier SETI programs has now been put in place. The amateur programs are adding valuable backup to the professionals, and valued moral and financial support. However, the professional programs are living hand-to-mouth, and have only survived through enlightened personal patronage. It should not have to be that way. SETI is a noble search based on sound scientific knowledge. The questions SETI is seeking to answer are surely among the most profound that science can address, and are of immense importance to the whole of humanity in understanding the meaning of life and our place in the cosmos.

The leadership of SETI rightly rests in the United States (where it all began), despite the willingness and enthusiasm of scientists in other countries to lend their support. The financial and intellectual capability lies with the United States to lead SETI for the whole of humankind. International collaboration on SETI is reasonably healthy, and has been made possible through the ingenuity of individuals and the established science "networks." The International Astronomical Union does have a commission for bioastronomy, and this does keep a paternalistic eye on SETI, but a firm commitment to the goals of SETI from the governments of the leading scientific nations is needed. The nations particularly strong in astronomy and space science are the United States, the United Kingdom, the Netherlands, Germany, France, Russia, Australia, Italy, Canada, and Japan. Their commitments would need to be long term. Since we have argued that the incidence of ETI may be relatively rare (at most thousands in the Milky Way Galaxy, rather than the millions once postulated), surveys of current sophistication could take decades. Nevertheless, commitments to scientific surveys of long duration have been made in the past, and similar vision is now called for from political and scientific leaders. The challenge is too important for interest to wane so soon.

The step we have taken of arguing for an expansion of SETI to embrace a form of ufology is, we freely acknowledge, unconventional. Let us make it quite clear (again) that we certainly do not believe the majority of reports of sightings of UFOs, the claimed visits of aliens, and the stories of alien abductions fed to the gullible readership of the tabloid press. What is being reported is largely fantasy and fraud. Nor do we support any of the activities of the pseudoscience fringe, with their stories of government conspiracies and their bogus "science." They are perpetrators of intellectual fraud. However, hidden in the "noise" of fraud, erroneous sightings, and myth, there just might be phenomena and events worthy of serious scientific investigation. Unless genuine science is prepared to take a serious look at the possibility of visitations from aliens, then fantasy and fraud will prevail. The Condon report should not be the final word on the subject from science (despite its thoroughness and supportive reviews), nor should the few individuals who have given UFOs serious scientific consideration be left to work in isolation.

Our enthusiasm for SETI is based on the evidence from current scientific understanding of the nature of the cosmos and the evolution of life on Earth. The same scientific understanding that supports SETI can also be used to argue for a scientific ufology. A simple logic would argue that if an extraterrestrial life-form has evolved to the stage where it has the ability to transmit radio signals to other worlds, then the technological step to interstellar space travel (at least with robot probes) is not that great. Of course, the enormous distances to be traveled in the cosmos, the extraordinary energy requirements, and the extremely lengthy times involved may temper the desire of any ETI to undertake interstellar travel, but there is in principle no insurmountable technological barrier. Physics does not get in the way. And time is on ETI's side. For reasons that are far from obvious, science has stepped back from rational investigation of the possibility of alien visitations. We would like to see scientists think again about their present coyness to look for evidence of interstellar mobility. Despite the giggle factor and the wacky approach to UFOs used by the tabloid press and the pseudoscience fringe, if we accept the scientific case for SETI, it would be wrong not to try to establish a comparably sound scientific case for a scientifically legitimate ufology. If science is not prepared to contemplate a soundly based ufology, then the battle for truth will have been lost to the misinformed cranks who currently peddle untruths and fantasies about UFOs, and those honest citizens who have seen things that they find difficult to understand will not be proffered sensible scientific explanations. If science does not take the trouble to explain people's genuinely believed sightings of unexplained objects, then it is hardly surprising that conspiracy theories have proved to be so popular. Our proposed science-based ufology does not start with urban-based mythology, conspiracy theory, or sightings of flying saucers. It starts with the contention that ETI could have an interstellar mobility capability, and then asks what evidence should we look for. The basis is the same as SETI, which starts with the contention that ETI could have a radio communication capability and then mounts surveys to obtain the evidence.

What needs to come next? A firm long-term commitment to SETI and a serious consideration of a genuinely science-based ufology. However, what if centuries of search were to prove that we are the sole technological civilization in the cosmos, and that life elsewhere is merely "slime"? Such a result, however improbable, would be even more profound than if SETI were eventually to demonstrate that ETI is commonplace. Either way, we need to appreciate our true place in the cosmos, for only then can we hope to understand our destiny as "citizens of the cosmos."

Glossary:
Common Scientific Terms Relevant to the Search for ETI

"Short words are best and the old words when short are best of all."

Winston Churchill

A

absolute zero—the lowest temperature theoretically possible, at which the energy of atoms and molecules is minimal; equivalent to -273 degrees centigrade.

accelerator—a machine that accelerates electrons or protons to very high energies to probe the structure of the atomic nucleus and understand the fundamental forces of nature.

acid rain—when pollutants such as sulfur dioxide are released from power stations or other sources into the atmosphere, they can be washed from the atmosphere in rain, forming what is called acid rain.

adenine—one of the major bases of DNA.

albedo—the ratio of radiation reflected from a surface to that incident on the surface.

alpha particle—a radioactive particle emitted from certain unstable atomic nuclei; contains two protons and two neutrons, so is equivalent to a helium nucleus.

AM—amplitude modulation; the method of transmitting a signal on a carrier wave, by varying the amplitude of the carrier wave.

amino acid—a group of water-soluble organic compounds that include an "amino group" ($-NH_2$) attached to a carbon atom.

amplifier—a device that increases the strength of an electric signal.

amplitude—the size of a disturbance, for example, the maximum displacement of a wave from its mean value.

analogue signal—a signal that varies continuously in amplitude with time.

antenna—the part of a radio or television system through which signals are transmitted (transmitting antenna) or received (receiving antenna).

antimatter—subatomic particles that have the same mass as their normal particle counterparts, but the opposite of some other property. For example, a positron has the same mass as an electron, but carries a positive rather than negative charge.

artificial intelligence—computer programs that perform tasks that require intelligence when performed by humans.

astrobiology—the science of the emergence and evolution of life in the cosmos.

atmosphere—the mixture of gases enveloping a planet. In the case of Earth, the main constituents of the dry atmosphere are nitrogen, oxygen, argon, and carbon dioxide.

atom—the smallest component of an element that can exist and still retain the characteristic properties of the element.

aurora—the brilliant atmospheric lighting effects produced at high latitudes by a stream of energetic particles entering the atmosphere from the solar wind.

B

background radiation—low-intensity background radioactivity from cosmic rays, rocks, and other sources.

bandwidth—the difference between the highest and lowest frequencies a system can transmit.

BETA—Billion-channel Extra-Terrestrial Assay.

beta particle—a radioactive particle (an electron) emitted from certain unstable atomic nuclei.

big bang—the "event" some 15 billion years ago that marked the creation of space and the beginning of time.

binary star—two stars orbiting one another.

biosphere—the whole of the land, sea, and atmosphere inhabited by living organisms.

black hole—a compact cosmic object whose gravitational field is so strong that its "escape velocity" (that is, the velocity an object must have to escape from it) exceeds the velocity of light.

blueshift—the blueward Doppler shift of light from galaxies approaching ours.

brown dwarf—a cosmic object intermediate between a planet and a star, but not massive enough to initiate fusion.

C

carbohydrates—a group of organic compounds linking carbon atoms with a varying number of water molecules, the most common form being the sugars such as glucose and sucrose.

carbon cycle—the natural cycle in which carbon is transferred between plants, animals, soil, and the atmosphere.

carbon dating—measuring the ratio of the radioactive carbon-14 isotope to normal carbon-12 in a sample of organic material (for example, wood) to estimate its age.

carrier wave—an electromagnetic wave that is modulated by a signal for transmission.

catalyst—a substance that is used to speed up a chemical reaction, although it is not itself changed in the reaction.

cell—in biology, the structural and functional unit of all living organisms; in physics, a system of electrodes placed in an electrolyte to produce an electric current.

cell membrane—the outer wall of a living cell.

cell nucleus—the control center of a living cell; contains the genetic material.

cellulose—the main constituent of the cell walls of plants.

centigrade (Celsius) scale—the temperature scale that is defined by 0 degrees as the temperature of ice and water in equilibrium, and 100 degrees as the temperature of steam above boiling water.

chain reaction—the self-sustaining fission of heavy nuclei, stimulated by neutrons released in previous fissions. A runaway chain reaction produces an atomic explosion. A controlled chain reaction is used in a nuclear power station.

chaos—a mathematical technique to describe apparently unpredictable and random behavior.

charge—the electrical property of certain fundamental particles that causes them to interact; comes in two forms: "positive" and "negative."

chemical reaction—the change in which one or more compounds or elements interact to form one or more new compounds.

chlorofluorocarbons—compounds in which some of the hydrogen atoms in hydrocarbons are replaced by chlorine and fluorine atoms; formerly used as refrigerants in refrigerators and propellants in aerosols; now being phased out because of their contribution to the destruction of the ozone layer.

chromosome—thread-like structures in the nuclei of living cells that contain the genes that determine the hereditary characteristics of an organism.

climate—the characteristic pattern of weather (sunshine, rainfall, humidity, wind) in a region over time.

clone—a cell or organism that has arisen from asexual reproduction, and is therefore genetically identical to the cell or organism from which it was derived.

clusters—a grouping of stars or galaxies.

coherent radiation—radiation in which sets of waves add constructively; that is, their peaks and troughs are exactly aligned.

comet—a fragment of debris from the formation of the Solar System, orbiting the sun in a highly elliptical orbit. When it is close to the sun, material evaporating from its surface is swept back into a characteristic tail by the solar wind.

compass—a small magnet pivoted at its midpoint so that it is free to rotate under the influence of the Earth's magnetic field.

compound—a substance formed from the combination of different elements in a fixed proportion.

continental drift—the theory that the continents were once accumulated as two giant land masses, which then broke up, with the components moving away from each other. The drift is now recognized as being driven by plate tectonics.

corona—the hot (millions of degrees) outer regions of the Sun's atmosphere.

cosmic rays—energetic particles of cosmic origin hitting the Earth's atmosphere.

cosmology—the study of the origin, present state, and future fate of the universe.

cytoplasm—the jelly-like material surrounding the nucleus of a cell.

cytosine—one of the major bases of DNA.

D

Darwinism—the theory of evolution based on natural selection.

data—information that has been prepared for a specific purpose.

database—a large body of information coded and stored in a computer.

decay—in radioactivity, the spontaneous release of a particle from the nucleus of an atom.

decomposer—an organism that gains energy from animal or plant waste, and the chemical breakdown of dead organisms; examples are earthworms, certain bacteria, and fungi.

decomposition—a reaction in which a chemical compound breaks down into simpler compounds.

demodulation—extracting the signal from a modulated carrier wave.

deoxyribonucleic acid (DNA)—the double-helix molecule that is the genetic material of most living organisms.

digital computer—an electronic system used to perform calculations and process data coded in digital form.

digital signal—the coding of a signal as a sequence of numbers.

disintegration—a process in which an atomic nucleus breaks up spontaneously into two or more components.

DNA—see deoxyribonucleic acid.

Doppler effect—the shift in wavelength observed when a source of sound (or light) is moving relative to an observer. The detected wavelength is shortened if the source is moving toward the observer, and lengthened when receding.

double helix—the popular name for the DNA molecule, describing its characteristic structure.

Drake equation—an equation that lists the factors contributing to the likely incidence of ETI in the cosmos.

Dyson sphere—a shell of matter or space colonies around a star, to optimize the use of the star's energy.

E

eclipse—the obscuring of light from a celestial body when it passes behind another body. Thus, the sun is eclipsed when the moon passes in front of it when observed from the Earth's surface.

ecology—the study of the interaction of organisms with their environment.

ecosystem—the system of certain biological species and the physical environment with which they interact.

electromagnetic waves—waves generated by changing electric or magnetic fields; light, gamma, and X-rays, ultraviolet and infrared radiation, and radio waves and microwaves are all forms of electromagnetic waves.

electron microscope—a microscope that uses a beam of electrons rather than light to illuminate the object to be magnified.

electronics—the use of devices in electrical circuits to perform certain functions, such as amplification, manipulation of data, transmission of information, and so forth.

element—a substance that cannot be broken down into simpler substances by chemical means.

elementary particles—the assortment of fundamental particles making up everything in the universe.

elliptical galaxy—a type of galaxy with characteristic ellipsoidal form.

enzymes—proteins that act as catalysts in biochemical reactions.

equinox—either of the two occasions during the year when the sun appears to cross the celestial equator (either from South to North in the "vernal" equinox, or from North to South in the "autumnal" equinox).

escape velocity—the minimum velocity an object must acquire to escape from the gravitational field of a celestial body.

ETI—extraterrestrial intelligence.

evolution—the gradual process whereby the diversity of animals and plants evolves.

exobiology—one of the terms used to describe the study of extraterrestrial life.

exosphere—the outermost region of the atmosphere.

expansion of the universe—the expansion of the universe from the big bang epoch of creation of space and time some 15 billion years ago; observed by the recession of galaxies and the background radiation from the big bang.

F

fiber optics—fine glass fibers along which light signals can be transmitted.

fission—a breakup of certain heavy atomic nuclei stimulated by the capture of a neutron, with the release of energy and further neutrons; the basis of nuclear chain reactions.

flare—a violent ejection of hot gas from the surface of the Sun.

FM—frequency modulation; the method of placing a signal on a carrier wave, by varying the frequency of the carrier to reflect variations in the signal.

food chain—the transfer of energy from plants through a sequence of organisms, each of which eats the one below it in the chain.

fossil—the preserved remains of an organism that lived in the geological past.

fossil fuel—coal, oil, and natural gas; formed from the decaying remains of living organisms from the geological past.

frequency—the number of oscillations per second of an oscillating or vibrating object.

fungi—a group of organisms, formerly thought to be plants, but now classified as making up a distinct living kingdom.

fusion—the forcing together of light atomic nuclei, to form a heavier nucleus with the release of energy.

G

galaxy—a conglomerate of billions of stars, bound by gravitational attraction; found in characteristic spiral or elliptical forms but often take on an irregular configuration.

gamma rays—high-energy electromagnetic radiation, emitted from the nuclei of radioactive elements.

gene—a discrete heredity unit of a chromosome, consisting of DNA.

gene sequencing—determining the sequence of bases in a gene.

general theory of relativity—the theory developed by Einstein relating the effects of gravity to those of accelerated motion. A prediction of the theory was that light would be deflected in a gravitational field (confirmed experimentally).

genetic code—the means by which genetic information in DNA leads to the production of proteins. The code takes the form of triplets of bases in DNA.

genetic engineering—the technique whereby an organism is modified by inserting genes from another organism into its DNA.

genetics—the study of heredity.

genome—all the genes contained in a complete set of chromosomes.

gigahertz (GHz)—frequencies of billions of hertz.

global warming—a gradual increase in mean global temperature, believed to be caused by pollution of the atmosphere with greenhouse gases.

gravitation—the universal attractive force acting between all matter.

greenhouse effect—the absorption of infrared radiation from the Earth's surface in the atmosphere. Pollution is increasing the quantity of greenhouse gases in the atmosphere, increasing the heat trapped within the atmosphere and leading to global warming.

guanine—one of the major bases of DNA.

H

habitable zone—the region around a star where water-based life might develop.

habitat—the natural place of habitation for a certain species.

heredity—see genetics.

hertz (Hz)—an alternative to "cycles per second" as the unit of frequency.

I

ice ages—epochs when global temperature fell dramatically and regions of ice cover progressed toward the equator.

inert gas—the group of elements with valence shells that are full, and therefore do not easily undergo chemical reactions.

infection—an invasion of an organism by disease-bearing microorganisms.

infrared rays—electromagnetic radiation of wavelength slightly longer than visible light.

integrated circuit—the incorporation of a large number of electronic components on a single sample of semiconductor.

interference—the interaction of two or more waves. When the wave peaks coincide, then the interference is "constructive." When the wave peak coincides with a trough, then the interference is "destructive."

Internet—the global communication network for computers.

ionization—the process by which atoms lose or acquire electrons.

ionosphere—the region in the upper atmosphere that has been ionized by ultraviolet radiation from the Sun.

irradiation—the exposure of material to radioactive particles or ionizing electromagnetic radiation.

irregular galaxy—a conglomerate of stars, with an overall irregular shape rather than the ordered elliptical or spiral form seen in other galaxies.

J

jet stream—a strong easterly wind in the upper troposphere.

K

kingdom—categories into which organisms are classified.

L

laser—"light amplification by the stimulated emission of radiation"; a device for producing an intense coherent beam of electromagnetic radiation.

light—the form of electromagnetic radiation that can be detected by the human eye.

lightning—an electrical discharge between a charged cloud and the Earth's surface.

light-year—the distance a pulse of light travels in 1 year.

Local Group—the cluster of galaxies to which the Milky Way belongs.

lodestone—magnetite; a naturally occurring magnetic material.

M

magic frequency—a guess at what might be a favored signaling frequency for ETI.

magma—the molten material that originates from within the Earth's mantle; when cooled and solidified, it forms igneous rock.

magnet—a piece of material that generates a magnetic field.

magnetic field—a field of force that exists around a magnet, or a wire carrying an electric current, and that influences any other magnet or wire carrying a current lying within the field.

magnetosphere—the region in space defined by the Earth's magnetic field.

mantle—the hot molten region below the Earth's crust.

megahertz (MHz)—frequencies of millions of hertz.

meiosis—the type of cell division to form reproductive cells, where the cells formed in a meiosis division have half the number of chromosomes of the parent cell.

memory—those parts of a computer where program code and data are stored.

META—Megachannel Exra-Terrestrial Assay.

metabolism—the various chemical reactions that take place in a living organism.

meteor—fragments of cosmic debris burning up in the Earth's atmosphere (observed as a "shooting star").

meteorite—fragments of cosmic debris large enough to survive entry through the Earth's atmosphere; boulder size to tens of meters across.

microclimate—the climate conditions resulting from localized phenomena, for example, in a river valley.

microscope—a device for forming magnified images of small objects.

microwaves—electromagnetic waves of wavelength intermediate between infrared radiation and radio waves.

Milky Way—the conglomerate of 400 billion stars within which our Solar System lies.

mitosis—the division of a cell to form two daughter cells, each containing the full complement of chromosomes of the parent cell.

modulation—the process of placing a signal on a carrier wave for transmission.

Moho—the boundary between the Earth's crust and the mantle within.

molecule—the fundamental unit of a compound, made up of two or more atoms bonded together.

mutagen—an agent that causes defects in a cell by interfering with the DNA coding.

mutation—a random change in the genetic makeup of a cell.

N

natural selection—the evolution of species through the survival of the fittest and adaptation to environment and circumstances.

nebula—a gaseous cloud in the cosmos.

network—the linking together of computers through a communication network. It can be local to a single site or group of sites (a local area network, or LAN), or over a wide area (a wide area network, or WAN), or indeed global (the Internet).

neuron—cells that are the fundamental units of the nervous system. Sensory neurons transmit information to the central nervous system, while motor neurons pass information to effectors (muscles and glands).

neutrinos—elementary particles that have no charge, travel with the speed of light, and are (probably) massless.

neutron star—in the imploding core of a supernova (the event that heralds the death of a massive star), protons and electrons merge to become neutrons. The dense core made up almost entirely of neutrons spins rapidly, and may be detected as a pulsar.

nitrogen cycle—the natural cycle in which nitrogen is transferred between plants, animals, soil, and the atmosphere.

noble gases—see inert gases.

nova—a thermonuclear explosion on the surface of an evolved star (probably caused by transfer of matter from the atmosphere of a companion star).

nuclear energy—the energy that can be extracted from the nuclei of atoms, through either fission (the splitting of massive nuclei) or fusion (the merging of light nuclei).

nuclear fission—see fission.

nuclear forces—the short-range forces that hold the particles in an atomic nucleus together.

nuclear fusion—see fusion.

nucleic acid—a complex organic molecule in living systems; see deoxyribonucleic acid and ribonucleic acid.

O

occultation—the passage of one celestial body in front of another, as in the occultation of a star by the Moon.

orbit—the path of one object around another, whether bound by gravitational or some other force.

organic chemistry—the study of the compounds of carbon.

oscillation—a periodic vibration.

Ozma—the first radio search for ETI.

ozone—the molecular form of oxygen involving three oxygen atoms.

ozone layer—the layer of ozone that is formed in the upper atmosphere and filters out harmful ultraviolet radiation from the Sun.

P

paleoclimatology—the determination of past climate from the fossil record and other means.

pathogen—the general term for a disease-carrying microorganism.

PCM—pulse-code modulation; the conversion of an analogue signal to a digital form for transmission.

period—the time interval between adjacent crests of a wave, or the time to complete one cycle of a regularly repeating phenomenon.

photon—the quantum (fundamental packet) of electromagnetic radiation.

photosphere—the visible surface of the Sun.

photosynthesis—the process by which green plants trap light to drive a sequence of chemical reactions to produce carbohydrates and oxygen from water and carbon dioxide.

pitch—the characteristic of sound that describes its "highness" or "lowness" to a listener; related to frequency.

plasma—a highly ionized gas.

plate tectonics—the ponderous movement of large segments of the Earth's crust, accounting for the slow drift of continents, mountain formation, volcanic activity, and many other geophysical phenomena.

positron—a fundamental particle with the same mass as an electron, but carrying the positive unit of charge (the "antiparticle" of an electron).

program—the sequence of instructions to be performed by a computer.

protein—the large group of organic compounds (made up of carbon, oxygen, hydrogen, nitrogen, and often sulfur) found in living organisms; comprises long chains of amino acids.

protein sequencing—the process of determining the sequence of amino acids in proteins.

protoplasm—the living contents of a cell, but excluding material recently ingested or waiting excretion.

pulsar—a rapidly spinning neutron star, emitting a collimated beam of radio waves that are detected as a series of rapid pulses as the star rotates.

Q

quantum—a minimum quantity by which energy can change.

quantum theory—the theory devised by Planck whereby energy can only be emitted in discrete amounts called quanta (plural of quantum). The energy of quanta is given by the Planck constant times the frequency of the emitted energy.

quasar—highly luminous celestial objects at extreme redshifts (and therefore at extreme distances, so that they are observed at an early stage in the evolution of the universe); thought to be nascent galaxies.

R

radar—the method of detecting echoes of transmitted radio waves from distant objects, to establish their presence and calculate their distance.

radiation—the term used to describe both electromagnetic waves and radioactive particles.

radiation damage—the damage to living systems caused by exposure to excessive amounts of radiation.

radioactivity—the spontaneous emission of energetic particles (alpha particles and beta particles) and gamma rays from the nuclei of certain elements.

radio astronomy—the detection of radio waves from celestial objects.

radio waves—electromagnetic waves of the longest wavelength.

red giant—a large luminous star at an advanced stage of evolution.

redshift—the redward Doppler shift of light from galaxies receding from ours.

reflection—the return of all or part of a wave or beam of particles on encountering a boundary.

refraction—the bending of a wave as it passes obliquely from one medium to another.

relativity—theories relating to relative motion; see general theory of relativity and special relativity.

ribonucleic acid (RNA)—compound organic molecules found in cells. Messenger RNA carries genetic code transcribed from DNA to sites in the cell called ribosomes, where it is translated into protein composition.

RNA—see ribonucleic acid.

S

scattering—the process by which electromagnetic radiation is deflected by particles in the medium through which it is passing.

science—the human endeavor dedicated to understanding the nature of, and patterns of behavior in, everything around us, and to making predictions based on that understanding.

scintillation—the "twinkling" of stars, caused by turbulence in the atmosphere.

SETI—search for extraterrestrial intelligence.

shock wave—a narrow region of high pressure formed when a projectile passes through a fluid at faster than the speed of sound.

signal—the means by which information is transmitted.

solar activity—the variable nature of the Sun, as evidenced by the presence of sunspots, solar flares, and other features.

solar energy—the electromagnetic energy radiated by the Sun.

Solar System—the Sun and its system of nine planets, their moons, and interplanetary objects such as asteroids and comets.

solar wind—the flux of energetic particles radiated by the Sun.

special relativity—the theory developed by Einstein to describe the effects of relative motion, based on the proposition that the velocity of light is independent of the velocity of its source; led to predictions about time dilation, length contraction, and the equivalence of mass and energy.

spectrum—a range of electromagnetic radiations spread out according to their wavelength.

spiral galaxy—a conglomerate of stars, with the characteristic form of intertwined spiral arms (like a giant Katherine wheel).

star—a self-luminous celestial body, generating energy by nuclear fusion in a central core.

stratosphere—the region of the Earth's atmosphere directly above the troposphere (the lower region responsible for weather).

subatomic particles—the fundamental particles from which atoms are formed.

summer solstice—the point at which the Sun appears to reach its highest above the celestial equator; marks the longest day of the year.

sunspot—a dark region on the Sun's photosphere, defining a region of lower-than-average temperature. The incidence of sunspots waxes and wanes on an 11-year cycle.

supernova—the violent explosion of a massive star that has reached the end of its normal evolution.

T

taxonomy—the classification of living and extinct species.

technology—the devices, systems, and processes, derived from scientific knowledge and engineering practice, that contribute to our lifestyles in a useful way.

telescope—an instrument able to collect light from a faint distant object in order to produce a visible image of it; can utilize either a lens or mirror system (or a combination of both) to collect and focus the light.

thermonuclear reaction—see fusion.

thymine—one of the major bases of DNA.

tides—the regular rise and fall of the oceans caused by the gravitational influence of the Moon.

tissue—a collection of similar cells that perform a particular function.

toxin—a poison that affects a living organism, for example, that produced by a bacterium.

transmutation—the transformation of one element into another, by the bombardment of particles.

troposphere—the lowest region of the Earth's atmosphere, in which most of the weather occurs.

U

ultraviolet radiation—electromagnetic radiation with wavelengths just shorter than those of visible light.

uncertainty principle—the principle defined by Heisenberg, stating that it is not possible to define both the position and the momentum of an atomic particle simultaneously, since an attempt to measure one will perturb the other.

universe—everything that is known to exist.

upper atmosphere—the outer reaches of the Earth's atmosphere, above an altitude of about 300 kilometers.

V

Van Allen radiation belts—the belts of energetic particles trapped in the Earth's magnetic field.

variable star—a star that increases and decreases its size periodically, and changes in luminosity.

virus—a particle that is capable of independent metabolism and reproduction within a living cell, but is completely inert outside a host cell. Viral diseases include the common cold, influenza, polio, smallpox, and acquired immunodeficiency syndrome (AIDS), and may also be linked to the development of certain cancers.

visible spectrum—the wavelengths of electromagnetic radiation that can be detected by the human eye.

W

water cycle—the cycle by which water is circulated between the oceans, rivers and lakes, and the atmosphere, and used by living organisms.

water hole—a quiet region in the radio spectrum lying between the hydrogen (H) and hydroxyl (OH) emission lines; considered to be a likely range for ETI signals.

wave—a periodic disturbance in a medium (for example, a water wave) or in space (for example, light). A wave is characterized by its amplitude, its velocity, its period, and its frequency.

wavelength—the distance between adjacent crests in a wave.

weather—the prevailing atmospheric conditions (including temperature, humidity, pressure) and the consequent occurrence of wind, rain, snow, and so forth.

white dwarf—the residual object left after a star of comparable mass to our Sun has used all its nuclear fuel. It is believed that some 999 out of every 1,000 stars eventually become a white dwarf.

winter solstice—the point at which the Sun appears to reach its lowest below the celestial equator; marks the shortest day of the year.

X
X-ray astronomy—the detection of X-rays from celestial objects.

X-ray stars—star systems that emit X-rays.

X-rays—energetic electromagnetic waves, with wavelengths shorter than those of ultraviolet radiation; extensively used for medical applications.

Y
year—the time it takes for the Earth to complete one revolution of its orbit around the Sun.

Z
zenith—the point on the celestial sphere directly above the observer.

Bibliography and Web Sites

Here is a list of some good World Wide Web sites to visit.

SETI Institute	http://www.seti-inst.edu
SERENDIP	http://sag.www.ssl.berkeley.edu /serendip
BETA	http://mc.harvard.edu/seti
SETI League	http://www.setileague.org
Columbus Optical SETI	http://www.coseti.org
Planetary Society	http://www.planetary.org
NASA Origins	http://origins.stsci.edu

Here are some good books for further reading on the subject.

Is Anyone Out There?, by Frank Drake and Dava Sobel. New York: Dell Publishing, 1994. (A thoroughly readable account from one of the SETI pioneers.)

Are We Alone?, by Paul Davies. New York: Basic Books, 1995.

Extraterrestrial Intelligence, by Jean Heidmann. New York: Cambridge University Press, 1997.

Pale Blue Dot, by Carl Sagan. New York: Random House, 1994.

We Are Not Alone, by Walter Sullivan. New York: Plume Books, 1993.

Life in the Universe. *Scientific American*. Special issue. Volume 271, no. 4. October 1994.

Extraterrestrials. Where Are They?, by Ben Zuckerman and Michael Hart. New York: Cambridge University Press, 1995.

The Biological Universe, by Steven Dick. New York: Cambridge University Press, 1996. (The best reference book available on the subject.)

So You Want to Be a SETI Scientist?

"I only took the regular course...
...the different branches of Arithmetic—
Ambition, Distraction, Uglification and Derision."
Lewis Carroll, *Alice's Adventures in Wonderland*

We are often asked by young people what they should do to become a professional scientist or engineer. The first thing to ask them is, "Are you good at mathematics?" If the answer is no, then the advice is that they should consider some different career options. If the answer is yes, then it is worth talking further about opportunities in science and engineering. Quite simply, in all branches of science and engineering the most important skills required include the numerical rigor, the algebraic techniques, and the ordered approach to logic that mathematics provides. For a young person who is not comfortable with mathematics to aspire to become a professional scientist or engineer is as nonsensical as someone who is tone-deaf aspiring to become an opera singer.

After "arithmetic" comes "ambition." Science is not easy, but it is exciting and it is a wonderful career for those with the ambition to succeed. To face up to all the hard work in gaining the desired qualification does require ambition, and commitment. A 3- or 4-year undergraduate degree in an appropriate scientific or engineering subject is the starting point. The nature of knowledge is such that "breadth" and "depth" in education are equally important. Thus, someone studying chemistry might still be expected to pick up new mathematical techniques, acquire computing skills, and learn about topics at interdisciplinary boundaries during undergraduate studies.

For someone who wants to apply knowledge, rather than generate it, a first degree will usually suffice, although perhaps some postgraduate course work can provide a further broadening of knowledge appropriate for a particular career path. Thus, a biology graduate, for example, might take an extra diploma course in information technology, to enable him or her to work on molecular graphics software used by geneticists.

To become a research scientist, ambition needs to be matched with dedication, and patience. Doctoral-level research training is required, and this will usually take a further 3 to 5 years beyond a first degree. To gain a doctorate, a student must perform a piece of original scholarship, of quality such as to be worthy of publication in the open research literature. The hurdle is a high one, and many stumble before reaching it.

In the increasingly competitive world of science, even clearing the doctoral hurdle may not suffice for those who wish to pursue a career in research. A postdoctoral fellowship, often taken in another country, may add a further 2 to 3 years of advanced training before a scientist (by now approaching 30 years old) can hope to secure a research career appointment in academia or industry. Even at this late stage, many scientists have to forego their ambition of a life in research; there simply are not enough career appointments in research to meet the aspirations of all those who have completed doctoral and postdoctoral training. A young scientist can face many distractions, as he or she tries to secure a career in research.

Most research scientists of our acquaintance are ambitious professionals, committed to personal advancement in a fiercely competitive vocation. There can be "uglification" aplenty in science, which as a profession is as competitive as any. Although scientists are motivated by a love of research, and the excitement generated by the pursuit of the unknown, human emotions can dominate the discovery process. However, bearing in mind the protracted apprenticeship they had to undertake to pursue a career in research, it is hardly surprising that research scientists are such a competitive breed!

The more esoteric the science discipline, the more difficult it is to establish a research career. Thus, a subject like astronomy has few career openings, and those who even get a doctorate can expect an especially long postdoctoral apprenticeship before there is any hope of a career appointment. However, the next generation of astronomers must be found, and for those with the talent, ambition, and patience, the opportunities will arise. Within astronomy, opportunities to get involved in SETI are especially restricted. But a young Frank Drake proved what is possible—and the next generation will produce a Jill Tatler, a Stu Bowyer, and a Paul Horowitz.

Even if there are few professional openings for those interested in SETI, there are opportunities to get involved through the SETI Institute, the SETI League, the Planetary Society, and SETI@home. But avoid, like the plague, the crank organizations peddling ETI fantasies. Stick with real science, and real SETI.

Index

tritium, 62, 240
troposphere, 128
type I civilizations, 68, 69, 129
type II civilizations, 68, 69, 70, 73, 76, 96, 106, 110, 111, 129, 226-229
type III civilizations, 68, 69, 70, 73, 76, 96, 106, 110, 111, 126, 226-228

U

ufology, 188, 189, 191, 193, 195-197, 210, 221, 223, 224, 242, 245, 248, 250, 254, 255
UFOs, 187-199, 209, 221, 222, 223, 244-250, 255
ultraviolet radiation, 30, 124, 128-129, 138, 147, 152, 167, 168
unidentified flying objects, *See* UFOs
United Nations, 86
United States, 57, 63-67, 69-76, 78-85, 103, 215, 253-254
University of California, 80, 81
University of Chicago, 167
uracil, 160
uranium, 121, 136, 162
Uranus, 121

V

Van Allen radiation belts, 125
Venus, 41, 49, 121, 126, 132
Verschuur, Gerritt, 73
vertebrates, 163, 173, 174
Viking, 2-5
visible light, 30, 31, 58, 124
vitamin D, 129
VLBI (very large baseline interferometry), 30
volcanoes, 26, 119, 126, 133, 137-139, 142, 151, 164
von Neumann probes, 240

W

Walpole, Horace, 81
water hole, 61, 66, 79, 81, 96
water vapor, 126-128, 131, 151, 167
Watson, James, 159-160
wavelength, 30, 46
waves, 142-143
weather, 119, 128, 132-134
Wegener, Alfred, 136
Wells, H. G., 3
whales, 52, 53
white dwarf, 41
Wilkins, Maurice, 160
winter solstice, 135
woodpeckers, 94-95
Wow signal, 75, 89, 110, 249

X

"The X-Files", 2, 195, 219
X-rays, 30, 31, 124, 159, 160

Z

Zuckerman, Ben, 73

About the Authors

Dr. David Clark is a New Zealander who has lived and worked in England for the past 24 years. He is married to Suzanne; they have three sons and live in Oxford. He started his career as a telecommunication engineer. His research background has been in space technology and astronomy. He has published approximately 80 research papers in learned journals, as well as numerous articles for popular science and technology magazines. Until 1985 he led the Space Astronomy research team at the Rutherford Appleton Laboratory in Oxfordshire, before moving into science and technology administration. He is currently director of engineering and science at the Engineering and Physical Sciences Research Council, the largest of the U.K. research agencies. His previous books include the following:

The Historical Supernovae, written with F. R. Stephenson (Pergamon Press)
Applications of Historical Astronomical Records, written with F. R. Stephenson (Adam Hilger)
Superstars (McGraw-Hill, and Dent)
The Quest for SS433 (Viking Penguin)
The Cosmos from Space (Crown)

Andrew Clark is the son of David Clark. He is at present undertaking postgraduate research in the philosophy of science, which has included the scientific and philosophical aspects of attempts to make contact with extraterrestrial intelligence.